CliffsQuickReview®
Trigonometry

By David A. Kay, MS

HOUGHTON MIFFLIN HARCOURT
BOSTON NEW YORK

About the Author

David A. Kay has been teaching Mathematics and Computer Information Systems at the college level for over 33 years. In addition, he has been working as a test prep instructor for the last 24 years.

Publisher's Acknowledgments

Editorial

Project Editor: Kelly Ewing
Senior Acquisitions Editor: Greg Tubach
Technical Editor: David Herzog
Editorial Administrator: Michelle Hacker

Composition

Indexer: TECHBOOKS Production Services
Proofreader: Arielle Carole Mennelle
Wiley Indianapolis Composition Services

CliffsQuickReview™ Trigonometry
Published by:
Houghton Mifflin Harcourt
www.hmhco.com

Copyright © 2001 Houghton Mifflin Harcourt
Library of Congress Control Number: 2001039155
ISBN: 978-0-7645-6389-8 (pbk); ISBN 978-1-118-15368-0 (ebk); ISBN 978-1-118-15418-2 (ebk);
ISBN 978-0-7645-2393-9 (ebk)
Printed in the United States of America
DOH 20 19 18 17 16 15
4500721463
1O/RV/QX/QU/IN
Published by Houghton Mifflin Harcourt
Published simultaneously in Canada

For information about permission to reproduce selections from this book, write to trade.permissions@hmhco.com or to Permissions, Houghton Mifflin Harcourt Publishing Company, 3 Park Avenue, 19th Floor, New York, New York 10016.

Table of Contents

INTRODUCTION

The word *trigonometry* comes from Greek words meaning measurement of triangles. Solving triangles is one of many aspects of trigonometry that you study today. To develop methods to solve triangles, trigonometric functions are constructed. The study of the properties of these functions and related applications form the subject matter of trigonometry. Trigonometry has applications in navigation, surveying, construction, and many other branches of science, including mathematics and physics.

Why You Need This Book

Can you answer yes to any of these questions?

- Do you need to review the fundamentals of trigonometry fast?
- Do you need a course supplement to trigonometry?
- Do you need a concise, comprehensive reference for trigonometry?

If so, then CliffsQuickReview Trigonometry is for you!

How to Use This Book

You're in charge here. You get to decide how to use this book. You can either read the book from cover to cover or just look for the information you need right now. However, here are a few recommended ways to search for topics:

- Flip through the book looking for your topics in the running heads.
- Look in the Glossary for all the important terms and definitions.
- Look for your topic in the Table of Contents in the front of the book.
- Look at the Chapter Check-In list at the beginning of each chapter.
- Look at the Chapter Check-Out questions at the end of each chapter.
- Test your knowledge with the CQR Review at the end of the book.

Visit Our Web Site

A great resource, www.cliffsnotes.com, features review materials, valuable Internet links, quizzes, and more to enhance your learning. The site also features timely articles and tips, plus downloadable versions of many CliffsNotes books.

When you stop by our site, don't hesitate to share your thoughts about this book or any Wiley product. Just click the Talk to Us button. We welcome your feedback!

Chapter 1

TRIGONOMETRIC FUNCTIONS

Chapter Checkin

- ☐ Understanding angles and angle measurements
- ☐ Finding out about trigonometric functions of acute angles
- ☐ Defining trigonometric functions of general angles
- ☐ Using inverse notation and linear interpolation

Historically, trigonometry was developed to help find the measurements in triangles as an aid in navigation and surveying. Recently, trigonometry is used in numerous sciences to help explain natural phenomena. In this chapter, I define angle measure and basic trigonometric relationships and introduce the use of inverse trigonometric functions.

Angles

An **angle** is a measure of rotation. Angles are measured in **degrees.** One complete rotation is measured as 360°. Angle measure can be positive or negative, depending on the direction of rotation. The angle measure is the amount of rotation between the two rays forming the angle. Rotation is measured from the **initial side** to the **terminal side** of the angle. Positive angles (Figure 1-1a) result from *counterclockwise* rotation, and negative angles (Figure 1-1b) result from *clockwise* rotation. An angle with its initial side on the x-axis is said to be in **standard position**.

Figure 1-1 (a) A positive angle and (b) a negative angle.

Angles that are in standard position are said to be **quadrantal** if their terminal side coincides with a coordinate axis. Angles in standard position that are not quadrantal fall in one of the four quadrants, as shown in Figure 1-2.

Figure 1-2 Types of angles.

Third quadrant angle
(a)

Second quadrant angle
(b)

First quadrant angle
(c)

Fourth quadrant angle
(d)

Quadrant angle
(e)

Quadrant angle
(f)

Example 1: The following angles (standard position) terminate in the listed quadrant.

94°	2nd quadrant
500°	2nd quadrant
−100°	3rd quadrant
180°	quadrantal
−300°	1st quadrant

Two angles in standard position that share a common terminal side are said to be coterminal. The angles in Figure 1-3 are all coterminal with an angle that measures 30°.

All angles that are coterminal with $d°$ can be written as

$$d° + n \cdot 360°$$

where n is an integer (positive, negative, or zero).

Example 2: Is an angle measuring 200° coterminal with an angle measuring 940°?

If an angle measuring 940° and an angle measuring 200° were coterminal, then

$$940° = 200° + n \cdot 360°$$

$$740° = n \cdot 360°$$

Because 740 is not a multiple of 360, these angles are not coterminal.

Figure 1-3 Angles coterminal with –30°.

(a)

(a)

(c)

(d)

(e)

(f)

Example 3: Name 4 angles that are coterminal with $-70°$.

$$-70° + (1)360° = 290°$$

$$-70° + (2)360° = 650°$$

$$-70° + (3)360° = 1110°$$

$$-70° + (-1)360° = -430°$$

$$-70° + (-2)360° = -790°$$

Angle measurements are not always whole numbers. Fractional degree measure can be expressed either as a decimal part of a degree, such as 34.25°, or by using standard divisions of a degree called minutes and seconds. The following relationships exist between degrees, minutes, and seconds:

1 degree = 60 minutes

1 minute = 60 seconds

or

$$1° = 60'$$
$$1' = 60''$$

Example 4: Write $34°15'$ using decimal degrees.

$$34°15' = 34° + \frac{15°}{60}$$

$$= 34° + .25°$$

$$= 34.25°$$

Example 5: Write $12°18'44''$ using decimal degrees.

$$12°18'44'' = 12° + \frac{18°}{60} + \frac{44°}{3600}$$

$$\approx 12° + .3° + .012°$$

$$\approx 12.312°$$

Example 6: Write $81.293°$ using degrees, minutes, and seconds.

$$81.293° = 81° + (.293 \times 60)'$$

$$= 81° + 17.58'$$

$$= 81° + 17' + (.58 \times 60)''$$

$$\approx 81° + 17' + 35''$$

$$\approx 81°17'35''$$

Functions of Acute Angles

The characteristics of **similar triangles**, originally formulated by Euclid, are the building blocks of trigonometry. Euclid's theorems state if two angles of one triangle have the same measure as two angles of another triangle, then the two triangles are similar. Also, in similar triangles, angle measure and ratios of corresponding sides are preserved. Because all right triangles contain a 90° angle, all right triangles that contain another angle of equal measure must be similar. Therefore, the ratio of the corresponding sides of these triangles must be equal in value. These relationships lead to the **trigonometric ratios**. Lowercase Greek letters are usually used to name angle measures. It doesn't matter which letter is used, but two that are used quite often are alpha (α) and theta (θ).

Angles can be measured in one of two units: **degrees** or **radians**. The relationship between these two measures may be expressed as follows:

$$180° = \pi \, radians$$

$$1° = \frac{\pi}{180} \, radians$$

$$1 \, radian = \frac{180°}{\pi}$$

The following ratios are defined using a circle with the equation $x^2 + y^2 = r^2$ and refer to Figure 1-4.

Figure 1-4 Reference triangles.

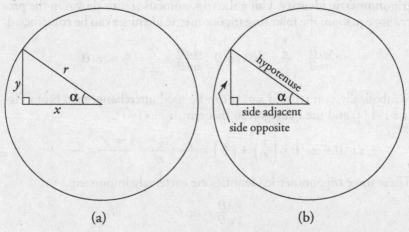

(a) (b)

$$sine \ of \ \alpha = \sin \alpha = \frac{y}{r} = \frac{length \ of \ side \ opposite \ \alpha}{length \ of \ hypotenuse}$$

$$cosine \ of \ \alpha = \cos \alpha = \frac{x}{r} = \frac{length \ of \ side \ adjacent \ to \ \alpha}{length \ of \ hypotenuse}$$

$$tangent \ of \ \alpha = \tan \alpha = \frac{y}{x} = \frac{length \ of \ side \ opposite \ \alpha}{length \ of \ side \ adjacent}$$

Remember, if the angles of a triangle remain the same, but the sides increase or decrease in length proportionally, these ratios remain the same. Therefore, trigonometric ratios in right triangles are dependent only on the size of the angles, not on the lengths of the sides.

The **cotangent, secant**, and **cosecant** are **trigonometric functions** that are the reciprocals of the **sine, cosine**, and **tangent**, respectively.

$$cosecant \ of \ \alpha = \csc \alpha = \frac{r}{y} = \frac{length \ of \ hypotenuse}{length \ of \ side \ opposite \ \alpha}$$

$$secant \ of \ \alpha = \sec \alpha = \frac{r}{x} = \frac{length \ of \ hypotenuse}{length \ of \ side \ adjacent \ to \ \alpha}$$

$$cotangent \ of \ \alpha = \cot \alpha = \frac{x}{y} = \frac{length \ of \ side \ adjacent \ to \ \alpha}{length \ of \ side \ opposite \ to \ \alpha}$$

If trigonometric functions of an angle θ are combined in an equation and the equation is valid for all values of θ, then the equation is known as a **trigonometric identity**. Using the trigonometric ratios shown in the preceding equation, the following trigonometric identities can be constructed.

$$\circ \frac{\sin \theta}{\cos \theta} = \frac{\frac{y}{r}}{\frac{x}{r}} = \frac{x}{y} = \tan \theta \quad \frac{\cos \theta}{\sin \theta} = \frac{\frac{x}{r}}{\frac{y}{r}} = \frac{x}{y} = \cot \theta$$

Symbolically, $(\sin \alpha)^2$ and $\sin^2 \alpha$ can be used interchangeably. From Figure 1-4 (a) and the Pythagorean theorem, $x^2 + y^2 = r^2$.

$$\sin^2 \theta + \cos^2 \theta = \left(\frac{y}{r}\right)^2 + \left(\frac{x}{r}\right)^2 = \frac{y^2}{r^2} + \frac{x^2}{r^2} = \frac{x^2 = y^2}{r^2} = \frac{r^2}{r^2} = 1$$

These three trigonometric identities are extremely important:

$$\frac{\sin \theta}{\cos \theta} = \tan \theta$$

$$\frac{\cos \theta}{\sin \theta} = \cot \theta$$

$$\sin^2 \theta + \cos^2 \theta = 1$$

Example 7: Find $\sin\theta$ and $\tan\theta$ if θ is an acute angle $\left(0° < \theta < 90°\right)$ and $\cos\theta = 1/4$.

$$\sin^2\theta + \cos^2\theta = 1 \qquad \tan\theta = \frac{\sin\theta}{\cos\theta}$$

$$\sin^2\theta + \left(\frac{1}{4}\right)^2 = 1$$

$$\sin^2\theta = 1 - \frac{1}{16} \qquad \tan\theta = \frac{\frac{\sqrt{15}}{4}}{\frac{1}{4}}$$

$$\sin^2\theta = \frac{15}{16} \qquad \tan\theta = \left(\frac{\sqrt{15}}{4}\right)\left(\frac{4}{1}\right) = \sqrt{15}$$

$$\sin\theta = \sqrt{\frac{15}{16}} = \frac{\sqrt{15}}{4}$$

Example 8: Find $\sin\theta$ and $\cos\theta$ if θ is an acute angle $\left(0° < \theta < 90°\right)$ and $\tan\theta = 6$.

If the tangent of an angle is 6, then the ratio of the side opposite the angle and the side adjacent to the angle is 6. Because all right triangles with this ratio are similar, the hypotenuse can be found by choosing 1 and 6 as the values of the two legs of the right triangle and then applying the Pythagorean theorem.

$$r^2 = x^2 + y^2$$

$$r^2 = 6^2 + 1^2$$

$$r^2 = 36 + 1$$

$$r^2 = 37$$

$$r = \sqrt{37}$$

$$\sin\theta = \frac{length\ of\ side\ opposite\ \theta}{hypotenuse} = \frac{6}{\sqrt{37}}$$

$$\cos\theta = \frac{length\ of\ side\ adjacent\ to\ \theta}{hypotenuse} = \frac{1}{\sqrt{37}}$$

Trigonometric functions come in three pairs that are referred to as **cofunctions**. The sine and cosine are cofunctions. The tangent and cotangent are cofunctions. The secant and cosecant are cofunctions. From right triangle XYZ, the following identities can be derived:

$$\sin X = \frac{x}{z} = \cos Y \qquad \sin Y = \frac{y}{z} = \cos X$$

$$\tan X = \frac{x}{y} = \cot Y \qquad \tan Y = \frac{y}{x} = \cot X$$

$$\sec X = \frac{z}{y} = \csc Y \qquad \sec Y = \frac{z}{x} = \csc X$$

Using Figure 1-5, observe that $\angle X$ and $\angle Y$ are complementary.

Figure 1-5 Reference triangles.

Thus, in general:

$$\sin\alpha = \cos\left(90° - \alpha\right) \qquad \cos\alpha = \sin\left(90° - \alpha\right)$$

$$\tan\alpha = \cot\left(90° - \alpha\right) \qquad \cot\alpha = \tan\left(90° - \alpha\right)$$

$$\sec\alpha = \csc\left(90° - \alpha\right) \qquad \csc\alpha = \sec\left(90° - \alpha\right)$$

Example 9: What are the values of the six trigonometric functions for angles that measure 30°, 45°, and 60° (see Figure 1-6 and Table 1-1).

Figure 1-6 Drawings for Example 9.

(a) (b)

Table 1-1 Trigonometric Ratios for 30°, 45°, and 60° Angles

θ	sin θ	csc θ	cos θ	sec θ	tan θ	cot θ
30°	$\frac{1}{2}$	2	$\frac{\sqrt{3}}{2}$	$\frac{2\sqrt{3}}{3}$	$\frac{\sqrt{3}}{3}$	$\sqrt{3}$
45°	$\frac{\sqrt{2}}{2}$	$\sqrt{2}$	$\frac{\sqrt{2}}{2}$	$\sqrt{2}$	1	1
60°	$\frac{\sqrt{3}}{2}$	$\frac{2\sqrt{3}}{3}$	$\frac{1}{2}$	2	$\sqrt{3}$	$\frac{\sqrt{3}}{3}$

Functions of General Angles

Acute angles in standard position are all in the first quadrant, and all of their trigonometric functions exist and are positive in value. This is not necessarily true of angles in general. Some of the six trigonometric functions of quadrantal angles are undefined, and some of the six trigonometric functions have negative values, depending on the size of the angle. Angles in standard position have their terminal side in or between one of

the four quadrants. Figure 1-7 shows a point A *(x, y)* located on the terminal side of angle θ with r as the distance AO. Note that r is always positive. Based on the figures,

$$\sin\theta = \frac{y}{r}$$

$$\csc\theta = \frac{r}{y}; y \neq 0$$

$$\cos\theta = \frac{x}{r}$$

$$\sec\theta = \frac{r}{x}; x \neq 0$$

$$\tan\theta = \frac{y}{x}; x \neq 0$$

$$\cot\theta = \frac{x}{y}; y \neq 0$$

Figure 1-7 Positive angles in various quadrants.

If angle θ is a quadrantal angle, then either *x* or *y* will be 0, yielding the undefined values if the denominator is zero. The sign, positive or negative, of the trigonometric functions depends on which quadrant this point *A* *(x, y)* is located in. Table 1-2 summarizes this information.

Table 1-2 Signs of Trig Functions in Various Quadrants

Function	*Quadrant*			
	I	*II*	*III*	*IV*
sin θ, csc θ	+	+	−	−
cos θ, sec θ	+	−	−	+
tan θ, cot θ	+	−	+	−

One way to remember which functions are positive and which are negative in the various quadrants is to remember a simple four-letter acronym, **ASTC**. This acronym can remind you that **A**ll are positive in quadrant **I**, the **S**ine is positive in quadrant **II**, the **T**angent is positive in quadrant **III**, and the **C**osine is positive in quadrant **IV**. This acronym could stand for **A**rizona **S**tate **T**eacher's **C**ollege, **A**ll **S**tudents **T**ake **C**lasses, or some other four-word expression that will help you remember the relationships.

Table 1-3 summarizes the values of the trigonometric functions of quadrantal angles. Note that undefined values result from division by 0.

Table 1-3 Values of Trig Functions for Various Quadrantal Angles

	sin θ	*cos* θ	*tan* θ	*cot* θ	*sec* θ	*csc* θ
0°	0	1	0	undefined	1	undefined
90°	1	0	undefined	0	undefined	1
180°	0	−1	0	undefined	−1	undefined
270°	−1	0	undefined	0	undefined	−1

The six trigonometric functions of angles that are not acute can be converted back to functions of acute angles. These acute angles are called the **reference angles** (see Table 1-4). The value of the function depends on the quadrant of the angle. If angle θ is in the second, third, or fourth quadrant, then the six trigonometric functions of θ can be converted to equivalent functions of an acute angle. Geometrically, if the angle is in quadrant

II, reflect about the *y*-axis. If the angle is in quadrant IV, reflect about the *x*-axis. If the angle is in quadrant III, rotate 180°. Keep in mind the sign of the functions during these conversions to the reference angle.

Table 1-4 Reference Angle Values in Various Quadrants

Function	Quadrant		
	II	**III**	**IV**
sin θ	sin (180° – θ)	–sin (θ – 180°)	–sin (360° – θ)
cos θ	–cos (180° – θ)	–cos (θ – 180°)	cos (360° – θ)
tan θ	–tan (180° – θ)	tan (θ – 180°)	–tan (360° – θ)
cot θ	–cot (180° – θ)	cot (θ – 180°)	–cot (360° – θ)
sec θ	–sec (180° – θ)	–sec (θ – 180°)	sec (360° – θ)
csc θ	csc (180° – θ)	–csc (θ – 180°)	–csc (360° – θ)

Example 10: Find the six trigonometric functions of an angle α that is in standard position and whose terminal side passes through the point (–5, 12).

From the Pythagorean theorem, the hypotenuse can be found. Then, the six trigonometric functions follow from the definitions (Figure 1-8).

Figure 1-8 Drawing for Example 10.

$$sin\,\alpha = \frac{12}{13}$$

$$r^2 = x^2 + y^2$$

$$cos\,\alpha = -\frac{5}{13}$$

$$r^2 = (-5)^2 + 12^2$$

$$tan\,\alpha = -\frac{12}{5}$$

$$r^2 = 25 + 144$$

$$r^2 = 169$$

$$cot\,\alpha = -\frac{5}{12}$$

$$r = \sqrt{169} = 13$$

$$sec\,\alpha = -\frac{13}{5}$$

$$csc\,\alpha = \frac{13}{12}$$

Example 11: If sin θ = 1/3, what is the value of the other five trigonometric functions if cos θ is negative?

Figure 1-9 Drawing for Example 11.

Because sin θ is positive and cos θ is negative, θ must be in the second quadrant. From the Pythagorean theorem,

$$x^2 + 1^2 = 3^2; x^2 + 1 = 9; x^2 = 8; x = \sqrt{8} = 2\sqrt{2}$$

and then it follows that

$$cos\,\theta = -\frac{2\sqrt{2}}{3}$$

$$tan\,\theta = -\frac{1}{2\sqrt{2}} = -\frac{\sqrt{2}}{4}$$

$$cot\,\theta = -2\sqrt{2}$$

$$sec\,\theta = -\frac{3}{2\sqrt{2}} = -\frac{3\sqrt{2}}{4}$$

$$csc\,\theta = 3$$

Example 12: What is the exact sine, cosine, and tangent of 330°?

Because 330° is in the fourth quadrant, sin 330° and tan 330° are negative and cos 330° is positive. The reference angle is 30°. Using the 30° – 60° – 90° triangle relationship, the ratios of the three sides are 1, 2, $\sqrt{3}$. Therefore,

$$\sin 30° = \frac{1}{2} \qquad\qquad \sin 330° = -\frac{1}{2}$$

$$\cos 30° = \frac{\sqrt{3}}{2} \quad \text{and} \quad \cos 330° = \frac{\sqrt{3}}{2}$$

$$\tan 30° = \frac{\sqrt{3}}{2} \qquad\qquad \tan 330° = -\frac{1}{\sqrt{3}} = -\frac{\sqrt{3}}{3}$$

Tables of Trigonometric Functions

Calculators and tables are used to determine values of trigonometric functions. Most scientific calculators have function buttons to find the sine, cosine, and tangent of angles. The size of the angle is entered in degree or radian measure, depending on the setting of the calculator. Degree measure will be used here unless specifically stated otherwise. When solving problems using trigonometric functions, either the angle is known and the value of the trigonometric function must be found, or the value of the trigonometric function is known and the angle must be found. These two processes are inverses of each other. Inverse notations are used to express the angle in terms of the value of the trigonometric function. The expression $\sin\theta = 0.4295$ can be written as $\theta = \text{Sin}^{-1} 0.4295$ or $\theta = \text{Arcsin} 0.4295$ and these two equations are both read as "theta equals Arcsin 0.4295." Sometimes the expression "inverse sine of 0.4295" is used. Some calculators have a button marked "arc," which is pressed prior to the function key to express "arc" functions. Arc functions are used to find the measure of the angle if the value of the trigonometric function is known. If tables are used instead of a calculator, the same table is used for either process. Note: The use of calculators or tables gives only approximate answers. Even so, an equal (=) sign is sometimes used instead of an approximate (\approx or \cong) sign.

Example 13: What is the sine of 48°?

$$\sin 48° \approx 0.7431448255 \approx 0.7431$$

Example 14: What angle has a cosine of 0.3912?

$$\text{Cos}^{-1} 0.3912 \approx 66.97081247° \approx 66.97°$$

Although a calculator can find trigonometric functions of fractional angle measure with ease, this may not be true if you must use a table to look up the values. Tables cannot list *all* angles. Therefore, approximation must be used to find values between those listed in the table. This method is known as **linear interpolation**. The assumption is made that differences in function values are directly proportional to the differences of the measures of the angles *over small intervals*. This is not really true, but yields a better answer than just using the closest value in the table. This method is illustrated in the following examples.

Example 15: Using linear interpolation, find tan 28.43° given that tan 28.40° = 0.5407 and tan 28.50° = 0.5430.

$$0.10 \left\{ 0.03 \left\{ \begin{matrix} \tan 28.40° \approx 0.5407 \\ \tan 28.43° \approx \, ? \\ \tan 28.50° \approx 0.5430 \end{matrix} \right\} x \right\} 0.0023 \, (x = \textit{the difference})$$

Set up a proportion using the variable x.

$$\frac{0.03}{0.10} \approx \frac{x}{0.0023}$$

$$(0.03)(0.0023) \approx 0.10x$$

$$0.000069 \approx 0.10x$$

$$x \approx 0.0007$$

Because x is the difference between tan 28.40° and tan 28.43°,

$$\tan 28.43 \approx 0.5407 + 0.0007 \approx 0.5414$$

Example 16: Find the first quadrant angle α where $\cos \alpha \approx 0.2622$, given that $\cos 74° \approx 0.2756$ and $\cos 75° \approx 0.2588$.

$$1.0 \left\{ x \left\{ \begin{matrix} \cos 74.0° \approx 0.2756 \\ \cos \alpha \approx 0.2622 \\ \cos 75.0° \approx 0.5588 \end{matrix} \right\} 0.0134 \right\} 0.0168 \, (x = \text{the difference})$$

Set up a proportion using the variable x.

$$\frac{x}{1.0} \approx \frac{0.0134}{0.0168}$$

$$0.0168x \approx 0.0134$$

$$x \approx 0.8$$

Therefore, $\alpha \approx 74.0° + 0.8° \approx 74.8°$.

An interesting approximation technique exists for finding the sine and tangent of angles that are less than 0.4 radians (approximately 23°). The sine and tangent of angles less than 0.4 radians are approximately equal to the angle measure. For example, using radian measure, $\sin 0.15 \approx 0.149$ and $\tan 0.15 \approx 0.151$.

Example 17: Find θ in Figure 1-10 without using trigonometry tables or a calculator to find the value of any trigonometric functions.

Figure 1-10 Drawing for Example 17.

Because $\sin \theta = 5/23 \approx 0.21739$, the size of the angle can be approximated as 0.217 radians, which is approximately 12.46°. In reality, the answer is closer to 0.219 radians, or 12.56°—quite close for an approximation. If the Pythagorean theorem is used to find the third side of the triangle, the process could also be used on the tangent.

$$5^2 + x^2 = 23^2$$

$$25 + x^2 = 529$$

$$x^2 = 504$$

$$x = \sqrt{504} \approx 22.45$$

$$\tan \theta = \frac{5}{22.45} \approx 0.223$$

Thus, $\theta \approx 0.223$ radians. Also, a close approximation.

Example 18: Find the measure of an acute angle α accurate to the nearest minute if $\tan \alpha = 0.8884$.

$$\alpha = \mathrm{Tan}^{-1} 0.8884$$

$$\alpha \approx 41.6179°$$

Using a calculator

$$\alpha \approx 41° + (0.6179)(60)'$$

$$\alpha \approx 41° + 37'$$

$$\alpha \approx 41° 37'$$

Chapter Checkout

Q&A

1. True or false: An angle of size −20° is coterminal with an angle of size 700°.

2. Write 34.603 using degrees, minutes, and seconds.

3. If 0° < θ < 90° and sin θ = $\frac{2}{3}$, find cos θ.

4. If 0° < θ < 90° and tan θ = $\frac{4}{3}$, find sin θ.

5. What is the exact cosine of 210°?

6. Find the measure of an angle to the nearest minute if its cosine is 0.678.

7. What angle has a tangent of 3.4?

Answers: 1. T **2.** 34°36'10.8" **3.** $\cos\theta = \frac{\sqrt{5}}{3}$ **4.** sin θ = 4/5 **5.** $-\frac{\sqrt{3}}{2}$ **6.** 47°19' **7.** 73.61°.

Chapter 2
TRIGONOMETRY OF TRIANGLES

Chapter Checkin

❑ Figuring out trigonometric ratios to find missing parts of right triangles

❑ Using the law of cosines to solve triangles

❑ Applying the law of sines to solve triangles

❑ Finding the area of triangles by using trigonometric functions

Triangles are made up of three line segments. They meet to form three angles. The sizes of the angles and the lengths of the sides are related to one another. If you know the size (length) of three out of the six parts of the triangle (at least one side must be included), you can find the sizes of the remaining sides and angles. If the triangle is a right triangle, you can use simple trigonometric ratios to find the missing parts. In a general triangle (acute or obtuse), you need to use other techniques, including the law of cosines and the law of sines. You can also find the area of triangles by using trigonometric ratios.

Solving Right Triangles

All triangles are made up of three sides and three angles. If the three angles of the triangle are labeled $\angle A, \angle B,$ and $\angle C$, then the three sides of the triangle should be labeled as a, b, and c. Figure 2-1 illustrates how lowercase letters are used to name the sides of the triangle that are opposite the angles named with corresponding uppercase letters. If any three of these six measurements are known (other than knowing the measures of the three angles), then you can calculate the values of the other three measurements. The process of finding the missing measurements is known as **solving the triangle**. If the triangle is a right triangle, then one of the angles is 90°.

Therefore, you can solve the right triangle if you are given the measures of two of the three sides or if you are given the measure of one side and one of the other two angles.

Figure 2-1 Drawing for Example 1.

(a) (b)

Example 1: Solve the right triangle shown in Figure 2-1(b) if $\angle B = 22°$ and $b=16$.

Because the three angles of a triangle must add up to $180°$, $\angle A = 90° - \angle B$. Thus, $\angle A = 68°$.

$$\sin B = \frac{b}{c} \qquad\qquad \tan B = \frac{b}{a}$$
$$\sin 22° = \frac{16}{c} \qquad\qquad \tan 22° = \frac{16}{a}$$
$$c = \frac{16}{\sin 22°} \qquad\qquad a = \frac{16}{\tan 22°}$$
$$c = \frac{16}{.3746} \qquad\qquad a = \frac{16}{.4040}$$
$$c = 42.71 \qquad\qquad a = 39.60$$

The following is an alternate way to solve for sides a and c:

$$\csc B = \frac{c}{b} \qquad\qquad \cot B = \frac{a}{b}$$
$$\csc 22° = \frac{c}{16} \qquad\qquad \cot 22° = \frac{a}{16}$$
$$c = 16\left(\csc 22°\right) \qquad\qquad a = 16\left(\cot 22°\right)$$
$$c = 16(2.669) \qquad\qquad a = 16(2.475)$$
$$c = 42.71 \qquad\qquad a = 39.60$$

This alternate solution may be easier because no division is involved.

Example 2: Solve the right triangle shown in Figure 2-1(b) if $b = 8$ and $a = 13$.

You can use the Pythagorean theorem to find the missing side, but trigonometric relationships are used instead. The two missing angle measurements will be found first and then the missing side.

$$\tan B = \frac{b}{a}$$

$$= \frac{8}{13}$$

$$= 0.6154$$

$$\angle B = 31.6°$$

$$\tan A = \frac{a}{b}$$

$$= \frac{13}{8}$$

$$= 1.625$$

$$\angle A = 58.4°$$

$$\sin B = \frac{b}{c}$$

$$\sin 31.6° = \frac{8}{c}$$

$$c = \frac{8}{\sin 31.6°}$$

$$c = \frac{8}{0.5239}$$

$$c = 15.27$$

In many applications, certain angles are referred to by special names. Two of these special names are **angle of elevation** and **angle of depression**. The examples shown in Figure 2-2 make use of these terms.

Figure 2-2 a) Angle of elevation and b) angle of depression.

Example 3: A large airplane (plane *A*) flying at 26,000 feet sights a smaller plane (plane *B*) traveling at an altitude of 24,000 feet. The angle of depression is 40°. What is the line of sight distance (*x*) between the two planes?

Figure 2-3 illustrates the conditions of this problem.

Figure 2-3 Drawing for Example 3.

From Figure 2-3, you can find the solution by using the sine of 40°:

$$\sin 40° = \frac{2000'}{x}$$
$$x = \frac{2000'}{\sin 40°}$$
$$x = \frac{2000'}{.6428}$$
$$x = 3111.4'$$

Example 4: A ladder must reach the top of a building. The base of the ladder will be 25' from the base of the building. The angle of elevation from the base of the ladder to the top of the building is 64°. Find the height of the building *(h)* and the length of the ladder *(m)*.

Figure 2-4 illustrates the conditions of this problem.

Figure 2-4 Drawing for Example 4.

$$\tan 64° = \frac{h}{25'} \qquad\qquad \cos 64° = \frac{25'}{m}$$
$$2.05 = \frac{h}{25'} \qquad\qquad .4384 = \frac{25'}{m}$$
$$h = (25')(2.05) \qquad\qquad m = \frac{25'}{.4384}$$
$$h = 51.25' \qquad\qquad m = 57.02'$$

Example 5: A woodcutter wants to determine the height of a tall tree. He stands at some distance from the tree and determines that the angle of elevation to the top of the tree is 40°. He moves 30' closer to the tree, and now the angle of elevation is 50°. If the woodcutter's eyes are 5' above the ground, how tall is the tree?

Figure 2-5 can help you visualize the problem.

Figure 2-5 Drawing for Example 5.

From the small right triangle and from the large right triangle, the following relationships are evident:

$$\cot 50° = \frac{y}{x} \qquad \tan 40° = \frac{x}{y + 30'}$$

$$.8391 = \frac{y}{x} \qquad .8391 = \frac{x}{y + 30'}$$

$$y = 0.8381x \qquad x = 0.8391(y + 30')$$

Substituting the first equation in the second yields:

$$x = 0.8391(0.8391x + 30')$$

$$x = 0.7041x + 25.17'$$

$$.2959x = 25.17'$$

$$x = 85.06'$$

Note that 5' must be added to the value of x to get the height of the tree, or 90.06' tall.

Example 6: Using Figure 2-6, find the length of sides x and y and the area of the large triangle.

Figure 2-6 Drawing for Example 6.

Because this is an isosceles triangle, and equal sides are opposite equal angles, the values of x and y are the same. If the triangle is divided into two right triangles, the base of each will be 6. Therefore,

$$\tan 70° = \frac{h}{6}$$

$$2.747 = \frac{h}{6}$$

$$\cos 70° = \frac{6}{x}$$

$$h = (6)(2.747)$$

$$0.342 = \frac{6}{x}$$

$$h = 16.48$$

$$x = \frac{6}{0.342}$$

$$area = \frac{bh}{2}$$

$$x = 17.54$$

$$= \frac{(12)(16.48)}{2}$$

$$= 98.88 \text{ sq units}$$

Law of Cosines

The previous section covered the solving of *right* triangles. In this section, and the next, you see formulas that can solve *any* triangle. If α, β, and γ are the angles of any (right, acute, or obtuse) triangle, and a, b, and c are the lengths of the three sides opposite α, β, and γ, respectively, then

$$a^2 = b^2 + c^2 - 2bc\cos\alpha$$

$$b^2 = a^2 + c^2 - 2ac\cos\beta$$

$$c^2 = a^2 + b^2 - 2ab\cos\gamma$$

These three formulas are called the **Law of Cosines**. Each follows from the distance formula and is illustrated in Figure 2-7.

Figure 2-7 Reference triangle for Law of Cosines.

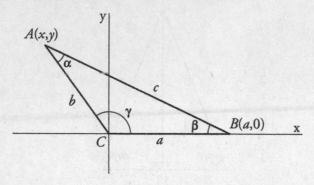

From the figure,

$$\sin\gamma = \frac{y}{b} \ and \ \cos\gamma = \frac{x}{b}$$

Thus the coordinates of A are

$$x = b\cos\gamma \ and \ y = b\sin\gamma$$

Remember, all three forms of the Law of Cosines are true even if γ is acute. Using the distance formula,

$$c^2 = \left(y-0\right)^2 + (x-a)^2$$
$$c^2 = \left(b\sin\gamma - 0\right)^2 + \left(b\cos\gamma - a\right)^2$$
$$c^2 = b^2\sin^2\gamma + b^2\cos^2\gamma - 2ab\cos\gamma + a^2$$
$$c^2 = b^2\left(\sin^2\gamma + \cos^2\gamma\right) - 2ab\cos\gamma + a^2$$
$$c^2 = a^2 + b^2 - 2ab\cos\gamma$$

In the preceding formula, if γ is 90°, then the cos 90° = 0, yielding the Pythagorean theorem for right triangles. If the orientation of the triangle is changed to have A or B at the origin, then the other two versions of the Law of Cosines can be obtained.

Two specific cases are of particular importance. First, use the Law of Cosines to solve a triangle if the length of the three sides is known.

Example 7: If α, β, and γ are the angles of a triangle, and a, b, and c are the lengths of the three sides opposite α, β, and γ, respectively, and $a = 12$, $b = 7$, and $c = 6$, then find the measure of β.

Use the form of the Law of Cosines that uses the angle in question.

$$b^2 = a^2 + c^2 - 2ac\cos\beta$$

Rewrite solving for cos β.

$$\cos\beta = \frac{a^2 + c^2 - b^2}{2ac}$$

$$\cos\beta = \frac{12^2 + 6^2 - 7^2}{(2)(12)(6)}$$

$$\cos\beta = \frac{131}{144} \approx 0.9097$$

Because cos β > 0 and β < 180°, β < 90°. Thus,

$$\beta \approx 24.54°$$

$$\approx 24°\,32'$$

The measure of α can be found in a similar way.

$$a^2 = b^2 + c^2 - 2bc\cos\alpha$$

Rewrite solving for cos α.

$$\cos\alpha = \frac{b^2 + c^2 - a^2}{2bc}$$

$$\cos\alpha = \frac{7^2 + 6^2 - 12^2}{(2)(7)(6)}$$

$$\cos\alpha = \frac{-59}{84} \approx -0.7024$$

Because cos α < 0 and α < 180°, α > 90°. Thus,

$$\alpha \approx 134.62°$$

$$\approx 134°\,37'$$

Because the three angles of the triangle must add up to 180°,

$$\gamma = 180° - \alpha - \beta$$

$$\gamma \approx 180° - 134.62° - 24.54°$$

$$\gamma \approx 20.84°$$

$$\approx 20°\,50'$$

Next, solve a triangle knowing the lengths of two sides and the measure of the included angle. First, find the length of the third side by using the Law of Cosines. Then proceed as in Example 7 to find the other two angles.

Example 8: Using Figure 2-8, find the length of side *b*.

Figure 2-8 Drawing for Example 8.

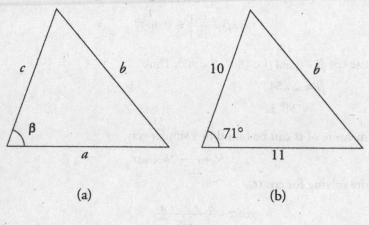

(a) (b)

$$b^2 = a^2 + c^2 - 2ac\cos\beta$$
$$b^2 = 11^2 + 10^2 - (2)(11)(10)\cos71°$$
$$b^2 \approx 121 + 100 - (220)(0.3256)$$
$$b^2 \approx 149.37$$
$$b \approx \sqrt{149.37}$$
$$b \approx 12.22$$

Example 9: Find the area of the triangle in Example 8.

First reposition the triangle as shown in Figure 2-9 so that the known angle is in standard position.

Figure 2-9 Drawing for Example 9.

The base of the triangle is 11. You can find the height of the triangle by using the fact that

$$\sin 71° = \frac{h}{10}$$

$$0.9455 \approx \frac{h}{10}$$

$$h \approx (10)(0.9455)$$

$$h \approx 9.455$$

Therefore,

$$area = \frac{base \times height}{2}$$

$$area = \frac{(11)(9.455)}{2}$$

$$area = 52 \; sq \; units$$

Law of Sines

You can use the Law of Cosines discussed in the last section to solve general triangles, but only under certain conditions. The formulas that will be developed in this section provide more flexibility in solving these general triangles.

The following discussion centers around Figure 2-10.

Figure 2-10 Reference triangles for Law of Sines.

(a) (b)

Line segment \overline{CD} is the altitude in each figure. Therefore $\triangle ACD$ and $\triangle BCD$ are right triangles. Thus,

$$\sin\alpha = \frac{h}{b} \Rightarrow h = b\sin\alpha$$

In Figure 2-10(b), $\angle CBD$ has the same measure as the reference angle for β. Thus,

$$\sin\beta = \frac{h}{a} \Rightarrow h = a\sin\beta$$

It follows that

$$b\sin\alpha = h = a\sin\beta$$

$$\frac{b\sin\alpha}{ab} = \frac{a\sin\beta}{ab}$$

$$\frac{\sin\alpha}{a} = \frac{\sin\beta}{b}$$

Similarly, if an altitude is drawn from A,

$$c\sin\alpha = h = a\sin\gamma$$

$$\frac{c\sin\alpha}{ac} = \frac{\alpha\sin\gamma}{ac}$$

$$\frac{\sin\alpha}{a} = \frac{\sin\gamma}{c}$$

Combining the preceding two results yields what is known as the **Law of Sines**.

$$\frac{\sin\alpha}{a} = \frac{\sin\beta}{b} = \frac{\sin\gamma}{c} \ or \ \frac{a}{\sin\alpha} = \frac{b}{\sin\beta} = \frac{c}{\sin\gamma}$$

In other words, in any given triangle, the ratio of the length of a side and the sine of the angle opposite that side is a constant. The Law of Sines is

valid for obtuse triangles as well as acute and right triangles, because the value of the sine is positive in both the first and second quadrant—that is, for angles less than 180°. You can use this relationship to solve triangles given the length of a side and the measure of two angles, or given the lengths of two sides and one opposite angle. (Remember that the Law of Cosines is used to solve triangles given other configurations of known sides and angles.)

First, consider using the Law of Sines to solve a triangle given two angles and one side.

Example 10: Solve the triangle in Figure 2-11 given $\theta = 32°$, $\phi = 77°$, and $d = 12$.

Figure 2-11 Drawing for Example 10.

It follows from the fact that there are 180° in any triangle that

$$\delta = 180° - \phi - \theta$$
$$\delta = 180° - 77° - 32°$$
$$\delta = 71°$$

From the Law of Sines, the following relationships are obtained:

$$\frac{d}{\sin\theta} = \frac{e}{\sin\delta} = \frac{f}{\sin\phi}$$
$$\frac{12}{\sin 32°} = \frac{e}{\sin 71°} = \frac{f}{\sin 77°}$$

Solving as two independent proportions,

$$\frac{12}{\sin 32°} = \frac{e}{\sin 71°}$$

$$\frac{12}{0.5299} \approx \frac{e}{0.9455}$$

$$e \approx \frac{(12)(0.9455)}{0.5299}$$

$$e \approx 21.41$$

$$\frac{12}{\sin 32°} = \frac{f}{\sin 77°}$$

$$\frac{12}{0.5299} \approx \frac{f}{0.9744}$$

$$f \approx \frac{(12)(0.9744)}{0.5299}$$

$$f \approx 22.07$$

Example 11: Solve the triangle in Figure 2-12 given $\alpha = 125°$, $\beta = 35°$, and $b, = 42$.

Figure 2-12 Drawing for Example 11.

From the fact that there are 180° in any triangle, then

$$\gamma = 180° - \alpha - \beta$$

$$\gamma = 180° - 125° - 35°$$

$$\gamma = 20°$$

Again, using the Law of Sines,

$$\frac{42}{\sin 35°} = \frac{c}{\sin 20°}$$

$$\frac{42}{0.5736} \approx \frac{c}{0.3420}$$

$$c \approx \frac{(42)(0.3420)}{0.5736}$$

$$c \approx 25.04$$

$$\frac{42}{\sin 35°} = \frac{a}{\sin 125°}$$

$$\frac{42}{0.5736} \approx \frac{a}{0.8192}$$

$$a \approx \frac{(42)(0.8192)}{0.5736}$$

$$a \approx 59.98$$

The second use of the Law of Sines is for solving a triangle given the lengths of two sides and the measure of the angle opposite one of them. In this case, it is possible that more than one solution will exist, depending on the values of the given parts of the triangle. The next example illustrates just such a case.

Example 12: Solve the triangle(s) given $a = 13$, $b = 20$, and $\alpha = 35°$.

Figure 2-13 shows two possible positions for side a. This will occur if $a > h$. To determine if a is greater than h, the value of h is found.

Figure 2-13 Drawing for Example 12.

$$\sin \alpha = \frac{h}{b}$$

$$\sin 35° = \frac{h}{20}$$

$$h \approx (20)(0.5736)$$

$$h \approx 11.47$$

Therefore, $a > h$ and the diagram shows two solutions for the triangle. From the Law of Sines,

$$\frac{\sin\alpha}{a} = \frac{\sin\beta}{b}$$

$$\frac{\sin 35°}{13} = \frac{\sin\beta}{20}$$

$$\sin\beta \approx \frac{(20)(0.5736)}{13}$$

$$\sin\beta = 0.8825$$

$$\beta \approx Sin^{-1} 0.8825$$

$$\beta \approx 61.95°$$

Because sin β is positive in both quadrant I and quadrant II, β can have two values and therefore two different solutions for the triangle.

$$\beta \approx 61.95°$$

$$\beta \approx 180° - 61.95° \approx 118.05°$$

Solution 1: The sum of the angles of a triangle is 180°. Thus,

$$\gamma = 180° - \alpha - \beta$$

$$\gamma \approx 180° - 35° - 61.95°$$

$$\gamma \approx 83.05°$$

Using the Law of Sines,

$$\frac{\sin\alpha}{a} = \frac{\sin\gamma}{c}$$

$$c = \frac{a\sin\gamma}{\sin\alpha}$$

$$c \approx \frac{(13)(\sin 83.05°)}{\sin 35°}$$

$$c \approx \frac{(13)(0.9927)}{0.5736}$$

$$c \approx 22.50$$

Solution 2: The sum of the angles of a triangle is 180°. Thus,

$$\gamma' = 180° - \alpha - \beta'$$

$$\gamma' = 180° - 35° - 118.05°$$

$$\gamma' = 26.95°$$

Using the Law of Sines,

$$\frac{\sin\alpha}{a} = \frac{\sin\gamma'}{c'}$$

$$c' = \frac{a\sin\gamma'}{\sin\alpha}$$

$$c' \approx \frac{(13)\left(\sin 26.95°\right)}{\sin 35°}$$

$$c' \approx \frac{(13)(0.4532)}{0.5736}$$

$$c' \approx 10.27$$

Solving General Triangles

The process of solving triangles can be categorized into several distinct groups. The following is a listing of these categories along with a procedure to follow to solve for the missing parts of the triangle. The assumption is made that all three missing parts are to be found. If only some of the unknown values are to be determined, a modified approach may be in order.

■ **SSS:** If the three sides of a triangle are known, first use the Law of Cosines to find one of the angles. It is usually best to find the largest angle first, the one opposite the longest side. Then, set up a proportion using the Law of Sines to find the second angle. Finally, subtract these angle measures from 180° to find the third angle.

The reason that the Law of Cosines should be used to find the largest angle in the triangle is that if the cosine is positive, the angle is acute. If the cosine is negative, the angle is obtuse. If the cosine is zero, the angle is a right angle. Once the largest angle of the triangle is known, the other two angles must be acute.

If the largest angle is not found by using the Law of Cosines but by using the Law of Sines instead, the determination whether the angle is acute or obtuse must be done using the Pythagorean theorem or other means because the sine is positive for both acute (first quadrant) and obtuse (second quadrant) angles. This adds an extra step to the solution of the problem.

If the size of only one of the angles is needed, use the Law of Cosines. The Law of Cosines may be used to find all the missing angles, although a solution using the Law of Cosines is usually more complex than one using the Law of Sines.

■ **SAS:** If two sides and the included angle of a triangle are known, first use the Law of Cosines to solve for the third side. Next, use the Law of Sines to find the smaller of the two remaining angles. This is the angle opposite the shortest or shorter side, not the longest side. Finally, subtract these angle measures from 180° to find the third angle. Again, you can use the Law of Cosines to find the two missing angles, although a solution using the Law of Cosines is usually more complex than one using the Law of Sines.

■ **ASA:** If two angles and the included side of a triangle are known, first subtract these angle measures from 180° to find the third angle. Next, use the Law of Sines to set up proportions to find the lengths of the two missing sides. You can use the Law of Cosines to find the length of the third side, but why bother if you can use the Law of Sines instead?

■ **AAS:** If two angles and a nonincluded side of a triangle are known, first subtract these angle measures from 180° to find the third angle. Next, use the Law of Sines to set up proportions to find the lengths of the two missing sides. The given side is opposite one of the two given angles. If all that is needed is the length of the side opposite the second given angle, then use the Law of Sines to calculate its value.

■ **SSA:** This is known as *the ambiguous case.* If two sides and a nonincluded angle of a triangle are known, there are *six* possible configurations, two if the given angle is obtuse or right and four if the given angle is acute. These six possibilities are shown in Figures 2-14, 2-15, and 2-16. In Figures 2-14 and 2-15, h is an altitude where $h = a \sin \beta$ and β is an acute angle.

In Figure 2-14(a), if $b < h$, then b cannot reach the other side of the triangle, and no solution is possible. This occurs when $b < a \sin \beta$.

In Figure 2-14(b), if $b = h = a \sin \beta$, then exactly one right triangle is formed.

Figure 2-14 Two cases for SSA.

(a) (b)

In Figure 2-15(a), if $h < b < a$—that is, $a \sin \beta < b < a$—then two different solutions exist.

In Figure 2-15(b), if $b \geq a$, then only one solution exists, and if $b = a$, then the solution is an isosceles triangle.

Figure 2-15 Ambiguous cases for SSA.

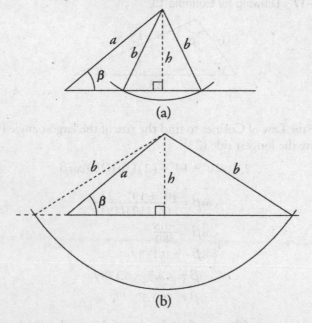

(a)

(b)

If β is an obtuse or right angle, the following two possibilities exist.

In Figure 2-16(a), if $b > a$, then one solution is possible.

In Figure 2-16(b), if $b \leq a$, then no solutions are possible.

Figure 2-16 Two cases for SSA.

(a) (b)

Example 13: (SSS) Find the difference between the largest and smallest angles of a triangle if the lengths of the sides are 10, 19, and 23, as shown in Figure 2-17.

Figure 2-17 Drawing for Example 13.

First, use the Law of Cosines to find the size of the largest angle (β) which is opposite the longest side (23).

$$23^2 = 10^2 + 19^2 - (2)(10)(19)\cos\beta$$

$$\cos\beta = \frac{10^2 + 19^2 - 23^2}{(2)(10)(19)}$$

$$\cos\beta = \frac{-68}{380}$$

$$\cos\beta = -0.1789$$

$$\beta \approx Cos^{-1} - 0.1789$$

$$\beta \approx 100.3°$$

Next, use the Law of Sines to find the size of the smallest angle (α), which is opposite the shortest side (10).

$$\frac{\sin\alpha}{10} \approx \frac{\sin 100.3^\circ}{23}$$

$$\sin\alpha \approx \frac{(10)(0.9839)}{23}$$

$$\sin\alpha \approx 0.4278$$

$$\alpha \approx Sin^{-1} - 0.4278$$

$$\alpha \approx 25.33^\circ$$

Thus, the difference between the largest and smallest angle is

$$100.3^\circ - 25.33^\circ = 74.97^\circ$$

Example 14: (SAS) The legs of an isosceles triangle have a length of 28 and form a 17° angle (Figure 2-18). What is the length of the third side of the triangle?

This is a direct application of the Law of Cosines.

$$c^2 = 28^2 + 28^2 - (2)(28)(28)\cos 17^\circ$$

$$c^2 \approx 784 - 784 - (1568)(0.9563)$$

$$c^2 \approx 68.52$$

$$c \approx \sqrt{68.52}$$

$$c \approx 8.278$$

Figure 2-18 Drawing for Example 14.

Example 15: (ASA) Find the value of *d* in Figure 2-19.

Figure 2-19 Drawing for Example 15.

First, calculate the sizes of angles α and β. Then find the value of *a* using the Law of Sines. Finally, use the definition of the sine to find the value of *d*.

$$\alpha = 180° - 65° = 115°$$
$$\beta = 180° - 115° - 31° = 34°$$
$$\frac{a}{\sin 115°} = \frac{120}{\sin 34°}$$
$$a = \frac{(120)(\sin 115°)}{\sin 34°}$$
$$a \approx \frac{(120)(0.9063)}{0.5592}$$
$$a \approx 194.5$$

Finally,

$$\sin 31° \approx \frac{d}{194.5}$$
$$d \approx (194.5)(0.5150)$$
$$d \approx 100.2$$

Example 16: (AAS) Find the value of *x* in Figure 2-20.

Figure 2-20 Drawing for Example 16.

First, calculate the size of angle α. Then use the Law of Sines to calculate the value of *x*.

$$\alpha = 180° - 110° - 22° = 48°$$

$$\frac{x}{\sin 48°} = \frac{41}{\sin 22°}$$

$$x \approx \frac{(41)\left(\sin 48°\right)}{\sin 22°}$$

$$x \approx \frac{(41)(0.7431)}{0.3746}$$

$$x \approx 81.33$$

Example 17: (SSA) One side of a triangle, of length 20, forms a 42° angle with a second side of the triangle (Figure 2-21). The length of the third side of the triangle is 14. Find the length of the second side.

Figure 2-21 Drawing for Example 17.

The length of the altitude *(h)* is calculated first so that the number of solutions (0, 1, or 2) can be determined.

$$h = (20)\left(\sin 42°\right) \approx (20)(0.6691) \approx 13.38$$

Because $13.38 < 14 < 20$, there are two distinct solutions.

Solution 1: Use of the Law of Sines to calculate α.

$$\frac{\sin \alpha}{20} = \frac{\sin 42°}{14}$$

$$\sin \alpha = \frac{(20)\left(\sin 42°\right)}{14}$$

$$\sin \alpha = \frac{(20)(0.6691)}{14}$$

$$\sin \alpha \approx 0.9559$$

$$\alpha \approx Sin^{-1} 0.9559$$

$$\alpha \approx 72.92°$$

Use the fact that there are 180° in a triangle to calculate β.

$$\beta = 180° - 42° - \alpha$$

$$\beta = 180° - 42° - 72.92°$$

$$\beta = 72.92°$$

Use the Law of Sines to find the value of *b*.

$$\frac{b}{\sin \beta} = \frac{14}{\sin 42°}$$

$$b = \frac{(14)\left(\sin \beta\right)}{\sin 42°}$$

$$b \approx \frac{(14)(0.9069)}{0.6691}$$

$$b \approx 18.98$$

Solution 2: Use α to find α', and α' to find β'

$$\alpha' = 180° - \alpha \approx 180° - 72.92° \approx 107.08°$$

$$\beta' = 180° - 42° - \alpha' \approx 180° - 42° - 107.08° \approx 30.92°$$

Next, use the Law of Sines to find b'.

$$\frac{b'}{\sin \beta'} = \frac{14}{\sin 42°}$$

$$b' = \frac{(14)(\sin \beta')}{\sin 42°}$$

$$b' = \frac{(14)(0.5138)}{0.6691}$$

$$b' \approx 10.75$$

Areas of Triangles

The most common formula for finding the area of a triangle is $K = \frac{1}{2} bh$, where K is the area of the triangle, b is the base of the triangle, and h is the height. (The letter K is used for the area of the triangle to avoid confusion when using the letter A to name an angle of a triangle.) Three additional categories of area formulas are useful.

Two sides and the included angle (SAS): Given $\triangle ABC$ (Figure 2-22), the height is given by $h = c \sin A$. Therefore,

$$K = \frac{1}{2} bh = \frac{1}{2} bc \sin A$$

$$K = \frac{1}{2} bh = \frac{1}{2} ac \sin B$$

$$K = \frac{1}{2} bh = \frac{1}{2} ab \sin C$$

Two angles and a side (AAS) or (ASA): Using the Law of Sines and substituting in the preceding three formulas leads to the following formulas:

$$\frac{c}{\sin C} = \frac{a}{\sin A}$$

$$c = \frac{a \sin C}{\sin A}$$

$$K = \frac{1}{2} ac \sin B$$

$$K = \frac{1}{2} a^2 \frac{\sin B \sin C}{\sin A}$$

Figure 2-22 Reference triangles for area formulas.

(a)

(b)

Similarly,

$$K = \frac{1}{2} b^2 \frac{\sin A \sin C}{\sin B}$$

$$K = \frac{1}{2} c^2 \frac{\sin A \sin B}{\sin C}$$

Three sides (SSS): A famous Greek philosopher and mathematician, Heron (or Hero), developed a formula that calculates the area of triangles given only the lengths of the three sides. This is known as **Heron's formula**. If *a, b,* and *c* are the lengths of the three sides of a triangle, and *s* is the **semiperimeter**, then

$$s = \frac{1}{2} (a + b + c)$$

and

$$K = \sqrt{s(s-a)(s-b)(s-c)}$$

One of many proofs of Heron's formula starts out with the Law of Cosines:

$$a^2 + b^2 - 2ab\cos C = c^2$$

$$a^2 + b^2 - c^2 = 2ab\cos C$$

$$\left(a^2 + b^2 - c^2\right)^2 = 4a^2 b^2 \cos^2 C$$

$$\left(a^2 + b^2 - c^2\right)^2 = 4a^2 b^2 \left(1 - \sin^2 C\right)$$

$$\left(a^2 + b^2 - c^2\right)^2 = 4a^2 b^2 - 4a^2 b^2 \sin^2 C$$

$$\left(a^2 + b^2 - c^2\right)^2 = 4a^2 b^2 - 4\left(ab\sin C\right)^2$$

$$\left(a^2 + b^2 - c^2\right)^2 = 4a^2 b^2 - 4\left(2K\right)^2$$

$$\left(a^2 + b^2 - c^2\right)^2 = 4a^2 b^2 - 16K^2$$

$$16K^2 = 4a^2 b^2 - \left(a^2 + b^2 - c^2\right)^2$$

$$16K^2 = \left[2ab + \left(a^2 + b^2 - c^2\right)\right]\left[2ab - \left(a^2 + b^2 - c^2\right)\right]$$

$$16K^2 = \left[(a+b)^2 + - c^2\right]\left[c^2 - (a-b)^2\right]$$

$$16K^2 = \left[(a+b) + c\right]\left[(a+b) - c\right]\left[c + (a-b)\right]\left[c - (a-b)\right]$$

$$16K^2 = (a+b+c)\left[(a+b+c) - 2c\right]$$

$$\times \left[(a+b+c) - 2b\right]\left[(a+b+c) - 2a\right]$$

$$16K^2 = (a+b+c)[2s - 2c][2s - 2b][2s - 2a]$$

$$16K^2 = (2s)\left[2(s-c)\right]\left[2(s-b)\right]\left[2(s-a)\right]$$

$$16K^2 = 16s(s-c)(s-b)(s-a)$$

$$K^2 = s(s-c)(s-b)(s-a)$$

$$K = \sqrt{s(s-a)(s-b)(s-c)}$$

Example 18: *(SAS)* As shown in Figure 2-23, two sides of a triangle have measures of 25 and 12. The measure of the included angle is 51°. Find the area of the triangle.

Figure 2-23 Drawing for Example 18.

Use the SAS formula:

$$K = \frac{1}{2} ab \sin C$$

$$K = \frac{1}{2}(25)(12) \sin 51°$$

$$K \approx \frac{1}{2}(25)(12)(0.7771)$$

$$K \approx 116.57 \text{ sq units}$$

Example 19: *(AAS and ASA)* Find the area of the triangle shown in Figure 2-24.

Figure 2-24 Drawing for Example 19.

First find the measure of the third angle of the triangle since all three angles are used in the area formula.

$$\angle C = 180° - 117° - 28° = 35°$$

$$K = \frac{1}{2} c^2 \frac{\sin A \sin B}{\sin C}$$

$$K = \frac{1}{2}(22)^2 \frac{\sin 117° \ \sin 28°}{\sin 35°}$$

$$K = \frac{1}{2}(22)^2 \frac{(0.8910)(0.4695)}{0.5736}$$

$$K = 176.5 \text{ sq units}$$

Example 20: *(AAS or ASA)* Find the area of an equilateral triangle with a perimeter of 78.

If the perimeter of an equilateral triangle is 78, then the measure of each side is 26. The nontrigonometric solution of this problem yields an answer of

$$K = \frac{s^2\sqrt{3}}{4} = \frac{26^2\sqrt{3}}{4} \approx \frac{676\sqrt{3}}{4} \approx 292.7 \text{ sq units}$$

The trigonometric solution yields the same answer.

$$K = \frac{1}{2}c^2 \frac{\sin A \ \sin B}{\sin C}$$

$$K = \frac{1}{2}(26)^2 \frac{\sin 60° \ \sin 60°}{\sin 60°}$$

$$K \approx \frac{1}{2}(26)^2 \frac{(0.8660)(0.8660)}{0.8660}$$

$$K \approx 292.7 \text{ sq units}$$

Example 21: *(SSS)* Find the area of a triangle if its sides measure 31, 44, and 60.

Use Heron's formula:

$$K = \sqrt{s(s-a)(s-b)(s-c)}$$

$$s = \frac{1}{2}(a+b+c)$$

$$s = \frac{1}{2}(31 + 44 + 60) = 67.5$$

$$K = \sqrt{(67.5)(67.5-31)(67.5-44)67.5-60}$$

$$K = \sqrt{(67.5)(36.5)(23.5)(7.5)}$$

$$K = \sqrt{434,235.9375}$$

$$K \approx 659 \text{ sq units}$$

Heron's formula does not use trigonometric functions directly, but trigonometric functions were used in the development and proof of the formula.

Chapter Checkout

Q&A

1. In a right triangle, the two legs have lengths of 8 inches and 10 inches. Find the size of the angle opposite the 10-inch side.

2. In a right triangle, the side opposite the 38° angle is 6.4 inches long. Find the length of the hypotenuse.

3. You are standing 400 feet from the base of a building. The angle elevation to the top of the building is 28°. Find the height of the building.

4. Find the size of the smallest angle in a triangle if the three sides measure 7 inches, 8 inches, and 10 inches.

5. Two sides of a triangle measure 12 feet and 18 feet, and the angle between them measures 16°. Find the length of the third side.

6. True or False: In some cases, the law of sines will not provide you with a unique answer.

7. Find the area of a triangle if its sides measure 10 feet, 14 feet, and 20 feet.

8. Find the area of a triangle if two of its sides measure 8 inches and 14 inches, and the angle between them measures 72°.

Answers: 1. 51.34° **2.** 10.395 inches **3.** 213 feet **4.** 44.05° **5.** 7.26 feet **6.** True **7.** 64.99 square feet **8.** 53.26 square inches

Chapter 3

GRAPHS OF TRIGONOMETRIC FUNCTIONS

Chapter Checkin

❑ Defining angles in terms of the radius of a circle

❑ Understanding the relationship between radian and degree measure

❑ Defining trig functions in terms of a unit circle

❑ Figuring out the period and symmetry of trig functions

❑ Creating the graphs of trig functions

You can measure angle sizes by using more than one scale. The *degree scale* is probably the most well-known scale, although the radian scale is equally as popular and useful. Although most applications deal only in one of these two scales, it is important to understand their differences and how to convert from one to the other.

You can define trig functions by using a unit circle, a circle with a radius of 1. As a point revolves around the circle, its distance from the x-axis is defined as the sine, and its distance from the y-axis is defined as the cosine. These definitions match the previous definitions in terms of a right triangle. The graphs of trigonometric functions are used to visually represent their behavior.

Radians

A central angle of a circle has an angle measure of 1° if it subtends an arc that is ¹⁄₃₆₀ of the circumference of the circle. This form of angle measure is quite common. Another form of angle measure that is in use is **radian measure**. If a central angle subtends an arc that is equal to the radius of the circle (Figure 3-1a), then the central angle has a measure of one radian.

Figure 3-1 Radian measure and subtended arcs.

(a) (b)

If a central angle θ of a circle with radius r subtends an arc of length q (Figure 3-1b), then its radian measure is defined as

$$\theta = \frac{q}{r} \ radians$$

Because both q and r are in the same units, when q is divided by r in the preceding formula, the units cancel. Therefore, *radian measure is unitless.*

Example 1: What is the radian measure of a central angle in a circle with radius 6 m if it subtends an arc of 24 m?

$$\theta = \frac{q}{r} = \frac{24m}{6m} = 4\,radians = 4\,rad = 4$$

(Note that if no units are listed for an angle measure, it is assumed to be in radians.)

If θ is one complete revolution, then the subtended arc is the circumference of the circle. In this case,

$$q = 2\pi r$$
$$\theta = \frac{2\pi r}{r}$$
$$\theta = 2\pi$$

Because one complete revolution is 360°,

$$360° = 2\pi$$
$$180° = \pi$$

The fact that 180° is the same as π radians is extremely important. From this relationship, the following proportion can be used to convert between radian measure and degree measure:

$$\frac{\theta°}{180°} = \frac{\theta \, rad}{\pi \, rad}$$

Example 2: What is the degree measure of 2.4 rad?

$$\frac{\theta}{180°} = \frac{2.4}{\pi}$$

$$\theta = \frac{(180)(2.4)}{\pi} = \frac{432°}{\pi}$$

$$\theta \approx 137.5°$$

Example 3: What is the radian measure of 63°?

$$\frac{63}{180} = \frac{\theta}{\pi}$$

$$\theta = \frac{(\pi)(63)}{180} = \frac{7\pi}{20} \, rad$$

$$\theta \approx 1.1 \, rad$$

The radian measures of many special angles follow directly from the radian-degree relationships. Some of these are summarized in Table 3-1.

Table 3-1 Degree/Radian Equivalencies

degrees	0°	30°	45°	60°	90°	120°	135°	150°	180°
radians	0	$\frac{\pi}{6}$	$\frac{\pi}{4}$	$\frac{\pi}{3}$	$\frac{\pi}{2}$	$\frac{2\pi}{3}$	$\frac{3\pi}{4}$	$\frac{5\pi}{6}$	π

The areas of sectors of a circle are directly proportional to the measures of their central angles and directly proportional to the arcs subtended by the central angles (Figure 3-2).

$$\frac{area \; of \; a \; sector \; of \; a \; circle}{area \; of \; the \; circle} = \frac{size \; of \; central \; angle \; of \; sector}{size \; of \; central \; angle \; of \; circle}$$

$$\frac{A}{\pi r^2} = \frac{\theta}{2\pi}$$

$$A = \frac{(\pi r^2)(\theta)}{2\pi}$$

$$A = \frac{r^2 \theta}{2}$$

Figure 3-2 Sector area.

Example 4: Find *r* given that A = 14π and θ = π/2.

$$\frac{A}{\pi r^2} = \frac{\theta}{2\pi}$$

$$\frac{14\pi}{\pi r^2} = \frac{\frac{\pi}{2}}{2\pi}$$

$$r^2 = \frac{(14\pi)(2\pi)}{\frac{\pi^2}{2}}$$

$$r^2 = 56$$

$$r = \sqrt{56}$$

$$r = 2\sqrt{14}$$

Example 5: Find θ if *A* = 6 and *r* = 4.

$$\frac{A}{\pi r^2} = \frac{\theta}{2\pi}$$

$$\frac{6}{\pi(4)^2} = \frac{\theta}{2\pi}$$

$$\theta = \frac{(6)(2\pi)}{16\pi}$$

$$\theta = \frac{3}{4}$$

Example 6: What is the angle measure, in radians, of the acute angle formed by the minute and hour hands of a clock at 7:15?

The hour hand moves ¹⁄₁₂ of a complete revolution each hour. Therefore, every 15 minutes (one quarter of an hour), the hour hand moves ¹⁄₄₈ of a complete revolution. Therefore, at 7:15, the hour and minute hands are ¹⁷⁄₄₈ of a revolution apart.

$$\frac{4}{12} + \frac{1}{48} = \frac{16}{48} + \frac{1}{48}$$

$\frac{17}{48}$ *of one complete revolution is*

$$\frac{17}{48} \times 2\pi = \frac{17\pi}{24} \text{ } radians \approx 127.5°$$

Example 7: Find the area of the shaded portion of the sector of the circle shown in Figure 3-3.

Figure 3-3 Drawing for Example 7.

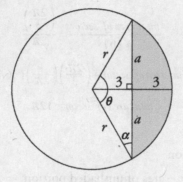

First, use the Pythagorean theorem to find the value of *a*.

$$3^2 + a^2 = r^2$$
$$3^2 + a^2 = 6^2$$
$$a^2 = 6^2 - 3^2$$
$$a^2 = 27$$
$$a = 3\sqrt{3}$$

The area of the triangular (unshaded) portion of the sector can be calculated using the area formula of a triangle.

$$\text{area of unshaded portion} = \frac{bh}{2} = \frac{(6\sqrt{3})(3)}{3} = 9\sqrt{3} \text{ sq units}$$

$$\sin\alpha = \frac{3}{6} = 0.5$$

$$\alpha = Sin^{-1} 0.5$$

$$\alpha = 30°$$

$$\alpha = \frac{\pi}{6} \ radians$$

It follows that

$$\theta = 120°$$

$$\theta = \frac{2\pi}{3} \ radians$$

$$\frac{total \ area \ of \ sector}{area \ of \ circle} = \frac{measure \ of \ central \ angle \ of \ sector}{measure \ of \ central \ angle \ of \ circle}$$

$$\frac{total \ area \ of \ sector}{\pi (6)^2} = \frac{\left(\frac{2\pi}{3}\right)}{2\pi}$$

$$total \ area \ of \ sector = \left(\frac{2\pi}{3}\right)\left(\frac{1}{2\pi}\right)(36\pi)$$

$$total \ area \ of \ sector = 12\pi$$

Therefore,

area of shaded portion

= area of total sector − area of unshaded portion

area of shaded portion = $12\pi - 9\sqrt{3} = 3\left(4\pi - 3\sqrt{3}\right)$

area of shaded portion ≈ 22.11

Circular Functions

The graph of the equation $x^2 + y^2 = 1$ is a circle in the rectangular coordinate system. This graph is called the **unit circle** and has its center at the origin and has a radius of 1 unit. Trigonometric functions are defined so that their *domains are sets of angles* and their ranges are sets of real numbers. **Circular functions** are defined such that their *domains are sets of numbers* that correspond to the measures (in radian units) of the angles of analogous trigonometric functions. The ranges of these circular functions, like their analogous trigonometric functions, are sets of real numbers. These functions are called circular functions because radian measures of angles are determined by the lengths of arcs of circles. In particular, trigonometric functions defined using the unit circle lead directly to these circular functions.

Begin with the unit circle $x^2 + y^2 = 1$ shown in Figure 3-4. Point A (1,0) is located at the intersection of the unit circle and the x-axis. Let q be any real number. Start at point A and measure $|q|$ units along the unit circle in a counterclockwise direction if $q \geq 0$ and in a clockwise direction if $q < 0$, ending up at point $P(x,y)$. Define the sine and cosine of q as the coordinates of point P. The other circular functions (the tangent, cotangent, secant, and cosecant) can be defined in terms of the sine and cosine.

$$\sin q = y$$

$$\cos q = x$$

$$\tan q = \frac{\sin q}{\cos q}; \cos q \neq 0$$

$$\cot q = \frac{\cos q}{\sin q}; \sin q \neq 0$$

$$\sec q = \frac{1}{\cos q}; \cos q \neq 0$$

$$\csc q = \frac{1}{\sin q}; \sin q \neq 0$$

Figure 3-4 Unit circle reference.

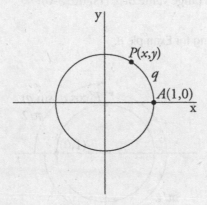

Sin q and cos q exist for each real number q because (cos q, sin q) are the coordinates of point P located on the unit circle, that corresponds to an arc length of $|q|$. Because this arc length can be positive (counterclockwise) or negative (clockwise), the *domain* of each of these circular functions is the set of real numbers. The *range* is more restricted. The cosine and sine are the abscissa and ordinate of a point that moves around the

unit circle, and they vary between -1 and 1. Therefore, the range of each of these functions is a set of real numbers z such that $-1 \leq z \leq 1$ (see Figure 3-5).

Figure 3-5 Range of values of trig functions.

Example 8: What value(s) x in the domain of the sine function between -2π and 2π have a range value of 1 (Figure 3-6)?

Figure 3-6 Drawing for Example 8.

The range value of $\sin x$ is 1 when point $P\left(\sqrt{8}/3, -1/3\right)$ has coordinates of (0, 1). This occurs when $x = \pi/2$ and $x = -3\pi/2$.

Example 9: What value(s) x in the domain of the cosine function between -2π and 2π have a range value of -1 (Figure 3-7)?

Figure 3-7 Drawing for Example 9.

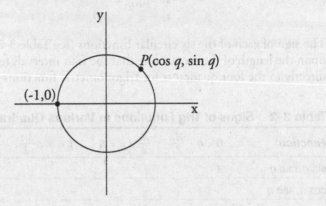

The range value of $\cos x$ is -1 when point $P(\cos x, \sin x)$ has coordinates of $(-1, 0)$. This occurs when $x = \pi$ and $x = -\pi$.

Example 10: The point $P(\sqrt{8}/3, -1/3)$ is on the unit circle. The length of the arc from point $A(1,0)$ to point P is q units. What are the values of the six circular functions of q?

The values of the sine and cosine follow from the definitions and are the coordinates of point P. The other four functions are derived using the sine and cosine.

$$\sin q = -\frac{1}{3}$$

$$\cos q = \frac{\sqrt{8}}{3}$$

$$\tan q = \frac{\sin q}{\cos q} = \frac{-\dfrac{1}{3}}{\dfrac{\sqrt{8}}{3}} = -\frac{1}{\sqrt{8}} = -\frac{\sqrt{2}}{4}$$

$$\cot q = \frac{\cos q}{\sin sq} = \frac{\dfrac{\sqrt{8}}{3}}{-\dfrac{1}{3}} = -\sqrt{8} = -2\sqrt{2}$$

$$\sec q = \frac{1}{\cos q} = \frac{1}{\sqrt{\frac{8}{3}}} = \frac{3}{\sqrt{8}} = \frac{3\sqrt{2}}{4}$$

$$\csc q = \frac{1}{\sin q} = \frac{1}{-\frac{1}{3}} = -3$$

The sign of each of the six circular functions (see Table 3-2) is dependent upon the length of the arc q. Note that the four intervals for q correspond directly to the four quadrants for trigonometric functions.

Table 3-2 Signs of Trig Functions in Various Quadrants

Function	$0 < q < \frac{\pi}{2}$	$\frac{\pi}{2} < q < \pi$	$\pi < q < \frac{3\pi}{2}$	$\frac{3\pi}{2} < q < 2$
sin q, csc q	+	+	−	−
cos q, sec q	+	−	−	+
tan q, cot q	+	−	+	−

Periodic and Symmetric Trigonometric Functions

The unit circle has a circumference of $\mathbf{C} = 2\pi r = 2\pi(1) = 2\pi$. Therefore, if a point P travels around the unit circle for a distance of 2π, it ends up where it started. In other words, for any given value q, if 2π is added or subtracted, the coordinates of point P remain unchanged (Figure 3-8).

Figure 3-8 Periodic coterminal angles.

It follows that

$$\sin(q + 2\pi) = \sin q$$

$$\sin(q - 2\pi) = \sin q$$

$$\cos(q + 2\pi) = \cos q$$

$$\cos(q - 2\pi) = \cos q$$

If k is an integer,

$$\sin(q - 2k\pi) = \sin q$$

$$\cos(q + 2\,k\,\pi) = \cos q$$

Functions that have this property are called **periodic functions**. A function f is periodic if there is a positive real number q such that $f(x + q) = f(x)$ for all x in the domain of f. The smallest possible value for q for which this is true is called the **period** of f.

Example 11: If $\sin y = \left(3\sqrt{5}\right)/10$, then what is the value of each of the following: $\sin\left(y + 8\pi\right)$, $\sin\left(y - 6\pi\right)$, $\sin\left(y + 210\pi\right)$?

All three have the same value of $\left(3\sqrt{5}\right)/10$ because the sine function is periodic and has a period of 2π.

The study of the periodic properties of circular functions leads to solutions of many real-world problems. These problems include planetary motion, sound waves, electric current generation, earthquake waves, and tide movements.

Example 12: The graph in Figure 3-9 represents a function f that has a period of 4. What would the graph look like for the interval $-10 \le x \le 10$?

Figure 3-9 Drawing for Example 12.

This graph covers an interval of 4 units. Because the period is given as 4, this graph represents one complete cycle of the function. Therefore, simply replicate the graph segment to the left and to the right (Figure 3-10).

Figure 3-10 Drawing for Example 12.

The appearance of the graph of a function and the properties of that function are very closely related. It can be seen from Figure 3-11 that

$$\cos (-q) = \cos q$$

$$\sin (-q) = -\sin q$$

Figure 3-11 Even and odd trig functions.

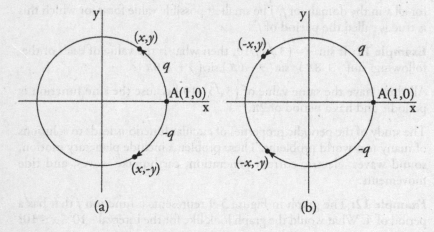

(a) (b)

The cosine is known as an **even function**, and the sine is known as an **odd function**. Generally speaking,

g is an even function if $g(-x) = g(x)$

g is an odd function if $g(-x) = -g(x)$

for every value of x in the domain of g. Some functions are odd, some are even, and some are neither odd nor even.

If a function is even, then the graph of the function will be symmetric with the y-axis. Alternatively, for every point (x, y) on the graph, the point $(-x, -y)$ will also be on the graph.

If a function is odd, then the graph of the function will be symmetric with the origin. Alternatively, for every point (x, y) on the graph, the point $(-x, -y)$ will also be on the graph.

Example 13: Graph several functions and give their periods (Figure 3-12).

Figure 3-12 Drawings for Example 13.

Example 14: Graph several odd functions and give their periods (Figure 3-13).

Figure 3-13 Drawings for Example 14.

Example 15: Is the function $f(x) = 2x^3 + x$ even, odd, or neither?

$$f(-x) = 2(-x)^3 + (-x)$$
$$= -2x^3 - x$$
$$= -(2x^3 + x)$$
$$= -f(x)$$

Because $f(-x) = -f(x)$, the function is odd.

Example 16: Is the function $f(x) = \sin x - \cos x$ even, odd, or neither?

$$f(-x) = \sin(-x) - \cos(-x)$$
$$= -\sin x - \cos x$$
$$= -(\sin x) + \cos x$$

Because $\quad -(\sin x + \cos x) \neq -(\sin x - \cos x)$

and $\quad -(\sin x + \cos x) \neq \sin x - \cos x$

the function is neither even nor odd. *Note: The sum of an odd function and an even function is neither even nor odd.*

Example 17: Is the function $f(x) = x \sin x \cos x$ even, odd, or neither?

$$f(-x) = (-x)\sin(-x)\cos(-x)$$
$$= (-x)(-\sin x)\cos x$$
$$= x\sin x\cos x$$
$$= f(x)$$

Because $f(-x) = f(x)$, the function is even.

Graphs of the Sine and Cosine

To see how the sine and cosine functions are graphed, use a calculator, a computer, or a set of trigonometry tables to determine the values of the sine and cosine functions for a number of different degree (or radian) measures (see Table 3-3).

Table 3-3 Values of the Sine and Cosine at Various Angles

degrees	0°	30°	45°	60°	90°	120°
radians	0	$\dfrac{\pi}{6}$	$\dfrac{\pi}{4}$	$\dfrac{\pi}{3}$	$\dfrac{\pi}{2}$	$\dfrac{2\pi}{3}$
sin x	0	0.500	0.707	0.866	1	0.866
cos x	1	0.866	0.707	0.500	0	−0.500

degrees	135°	150°	180°	210°	225°	240°
radians	$\dfrac{3\pi}{4}$	$\dfrac{5\pi}{6}$	π	$\dfrac{7\pi}{6}$	$\dfrac{5\pi}{4}$	$\dfrac{4\pi}{3}$
sin x	0.707	0.500	0	−0.500	−0.707	−0.866
cos x	−0.707	−0.866	−1	−0.866	−0.707	−0.500

degrees	270°	300°	315°	330°	360°
radians	$\dfrac{3\pi}{2}$	$\dfrac{5\pi}{3}$	$\dfrac{7\pi}{4}$	$\dfrac{11\pi}{6}$	2π
sin x	−1	−0.866	−0.707	−0.500	0
cos x	0	0.500	0.707	0.866	1

Next, plot these values and obtain the basic graphs of the sine and cosine function (Figure 3-14).

Figure 3-14 One period of the a) sine function and b) cosine function.

The sine function and the cosine function have periods of 2π; therefore, the patterns illustrated in Figure 3-14 are repeated to the left and right continuously (Figure 3-15).

Figure 3-15 Multiple periods of the a) sine function and b) cosine function.

(a)

(b)

Several additional terms and factors can be added to the sine and cosine functions, which modify their shapes.

The additional term A in the function $y = A + \sin x$ allows for a **vertical shift** in the graph of the sine functions. This also holds for the cosine function (Figure 3-16).

Figure 3-16 Examples of several vertical shifts of the sine function.

$y = 3 + \sin x$
$y = 2 + \sin x$
$y = 1 + \sin x$
$y = \sin x$
$y = -1 + \sin x$

The additional factor B in the function $y = B \sin x$ allows for **amplitude** variation of the sine function. The amplitude, $|B|$, is the maximum deviation from the x-axis—that is, one half the difference between the maximum and minimum values of the graph. This also holds for the cosine function (Figure 3-17).

Figure 3-17 Examples of several amplitudes of the sine function.

$y = 3 \sin x$
$y = 2 \sin x$
$y = \sin x$
$y = 1/2 \sin x$

Combining these figures yields the functions $y = A + B \sin x$ and also $y = A + B \cos x$. These two functions have **minimum** and **maximum** values as defined by the following formulas. The maximum value of the function is $M = A + |B|$. This maximum value occurs whenever $\sin x = 1$ or $\cos x = 1$. The minimum value of the function is $m = A - |B|$. This minimum occurs whenever $\sin x = -1$ or $\cos x = -1$.

Example 18: Graph the function $y = 1 + 2 \sin x$. What are the maximum and minimum values of the function?

The maximum value is $1 + 2 = 3$. The minimum value is $1 - 2 = -1$ (Figure 3-18).

Figure 3-18 Drawing for Example 18.

Example 19: Graph the function $y = 4 + 3 \sin x$. What are the maximum and minimum values of the function?

The maximum value is $4 + 3 = 7$. The minimum value is $4 - 3 = 1$ (Figure 3-19).

Figure 3-19 Drawing for Example 19.

The additional factor C in the function $y = \sin Cx$ allows for **period** variation (length of cycle) of the sine function. (This also holds for the cosine function.) The period of the function $y = \sin Cx$ is $2\pi/|C|$. Thus, the function $y = \sin 5x$ has a period of $2\pi/5$. Figure 3-20 illustrates additional examples.

Figure 3-20 Examples of several frequencies of the a) sine function and b) cosine function.

(a)

(b)

The additional term D in the function $y = \sin (x + D)$ allows for a **phase shift** (moving the graph to the left or right) in the graph of the sine functions. (This also holds for the cosine function.) The phase shift is $|D|$. This is a positive number. It does not matter whether the shift is to the left (if D is positive) or to the right (if D is negative). The sine function is odd, and the cosine function is even. The cosine function looks exactly like the sine function, except that it is shifted $\pi/2$ units to the left (Figure 3-21). In other words,

$$\sin x = \cos\left(x - \frac{\pi}{2}\right) \text{ or } \cos x = \sin\left(x + \frac{\pi}{2}\right)$$

Figure 3-21 Examples of several phase shifts of the sine function.

Example 20: What is the amplitude, period, phase shift, maximum, and minimum values of

1. $y = 3 + 2 \sin(3x - 2)$

2. $y = -2 + \frac{1}{2} \sin\left(\frac{1}{3}x + 2\right)$

3. $y = 4 \cos 2\pi x$

Table 3-4 Attributes of the General Sine Function

Function	Amplitude	Period	Phase Shift	Maximum	Minimum
$y = 3 + 2 \sin(3x - 2)$	2	$\frac{2\pi}{3}$	$2\,(\text{right})$	5	1
$y = -2 + \frac{1}{2} \sin\left(\frac{1}{3}x + 2\right)$	$\frac{1}{2}$	6π	$2\,(\text{right})$	$-\frac{3}{2}$	$-\frac{5}{2}$
$y = 4 \cos 2\pi x$	4	1	0	4	−4

Example 21: Sketch the graph of $y = \cos \pi x$.

Because $\cos x$ has a period of 2π, $\cos \pi x$ has a period of 2 (Figure 3-22).

Figure 3-22 Drawing for Example 21.

Example 22: Sketch the graph of $y = 3 \cos(2x + \pi/2)$.

Because $\cos x$ has a period of 2π, $\cos 2x$ has a period of π (Figure 3-23).

Figure 3-23 Drawing for Example 22.

The graph of the function $y = -f(x)$ is found by reflecting the graph of the function $y = f(x)$ about the x-axis. Thus, Figure 3-23 can also represent the graph of $y = -3 \sin 2x$. Specifically,

$$-\sin x = \sin (x \pm \pi)$$

$$-\cos x = \cos (x \pm \pi)$$

$$-\sin x = \cos (x + \pi/2)$$

$$-\cos x = \sin (x - \pi/2)$$

It is important to understand the relationships between the sine and cosine functions and how phase shifts can alter their graphs.

Graphs of Other Trigonometric Functions

The tangent is an odd function because

$$\tan(-x) = \frac{\sin(-x)}{\cos(-x)} = \frac{-\sin x}{\cos x} = -\tan x$$

The tangent has a period of π because

$$\tan(x + \pi) = \frac{\sin(x + \pi)}{\cos(x + \pi)} = \frac{-\sin x}{-\cos x} = \tan x$$

The tangent is undefined whenever $\cos x = 0$. This occurs when $x = q\pi/2$, where q is an odd integer. At these points, the value of the tangent approaches infinity and is undefined. When graphing the tangent, a dashed line is used to show where the value of the tangent is undefined. These lines are called **asymptotes**. The values of the tangent for various angle sizes are shown in Table 3-5.

Table 3-5 Values of the Tangent Function at Various Angles

degrees	0°	30°	45°	60°	75°	80°	85°	87°	90°
radians	0	$\dfrac{\pi}{6}$	$\dfrac{\pi}{4}$	$\dfrac{\pi}{3}$	$\dfrac{5\pi}{12}$	$\dfrac{4\pi}{9}$	$\dfrac{17\pi}{36}$	$\dfrac{29\pi}{60}$	$\dfrac{\pi}{2}$
tan x	0	0.577	1	1.73	3.73	5.67	11.43	19.08	*

* undefined

The graph of the tangent function over the interval from 0 to $\pi/2$ is as shown in Figure 3-24.

Figure 3-24 A portion of the tangent function.

The tangent is an odd function and is symmetric about the origin. The graph of the tangent over several periods is shown in Figure 3-25. Note that the asymptotes are shown as dashed lines, and the value of the tangent is undefined at these points.

Figure 3-25 Several periods of the tangent function.

The cotangent is the reciprocal of the tangent, and its graph is shown in Figure 3-26. Note the difference between the graph of the tangent and the cotangent in the interval from 0 to $\pi/2$.

Figure 3-26　A portion of the cotangent function.

As shown in Figure 3-27, in the graph of the cotangent, the asymptotes are located at multiples of π.

Figure 3-27　Several periods of the cotangent function.

Because the graphs of both the tangent and cotangent extend without bound both above and below the x-axis, the amplitude for the tangent and cotangent is not defined.

The general forms of the tangent and cotangent functions are

$$y = A + B \tan (Cx + D) \text{ and } y = A + B \cot (Cx + D)$$

The variables C and D determine the period and phase shift of the function as they did in the sine and cosine functions. The period is π/C and the phase shift is D/C. The shift is to the right if $D/C < 0$, and to the left if $D/C > 0$. The variable B does not represent an amplitude because the tangent and cotangent are unbounded, but it does represent how much the graph is "stretched" in a vertical direction. The variable A represents the vertical shift.

Example 23: Determine the period, phase shift, and the location of the asymptotes for the function

$$y = \tan\left(\frac{\pi}{3}x + \frac{\pi}{6}\right)$$

and graph at least two complete periods of the function.

The asymptotes can be found by solving $Cx + D = \pi/2$ and $Cx + D = -\pi/2$ for x.

$$\frac{\pi}{3}x + \frac{\pi}{6} = \frac{\pi}{2}$$
$$\frac{\pi}{3}x = \frac{\pi}{3}$$
$$x = 1$$

$$\frac{\pi}{3}x + \frac{\pi}{6} = -\frac{\pi}{2}$$
$$\frac{\pi}{3}x = -\frac{2\pi}{3}$$
$$x = -2$$

The period of the function is

$$\frac{\pi}{C} = \frac{\pi}{\left(\frac{\pi}{3}\right)} = 3$$

The phase shift of the function is

$$\frac{D}{C} = \frac{\left(\frac{\pi}{6}\right)}{\left(\frac{\pi}{3}\right)} = \frac{1}{2}$$

Because the phase shift is positive, it is to the left (Figure 3-28).

Figure 3-28 Phase shift of the tangent function.

$$y = \tan(\pi x/3 + \pi/6)$$

The amplitude is not defined for the secant or cosecant. The secant and cosecant are graphed as the reciprocals of the cosine and sine, respectively, and have the same period (2π). Therefore, the phase shift and period of these functions is found by solving the equations $Cx + D = 0$ and $Cx + D = 2\pi$ for x.

Example 24: Determine the period, phase shift, and the location of the asymptotes for the function

$$y = \frac{1}{2}\csc\left(\frac{\pi}{2}x + \frac{\pi}{2}\right)$$

and graph at least two periods of the function.

The asymptotes can be found by solving $Cx + D = 0$, $Cx + D = \pi$, and $Cx + D = 2\pi$ for x.

$$\frac{\pi}{2}x + \frac{\pi}{2} = 0$$
$$\frac{\pi}{2}x = -\frac{\pi}{2}$$
$$2x = -2$$
$$x = -1$$

$$\frac{\pi}{2}x + \frac{\pi}{2} = \pi$$

$$\frac{\pi}{2}x = \frac{\pi}{2}$$

$$\frac{x}{2} = \frac{1}{2}$$

$$x = 1$$

$$\frac{\pi}{2}x + \frac{\pi}{2} = 2\pi$$

$$\frac{\pi}{2}x = \frac{3\pi}{2}$$

$$\frac{x}{2} = \frac{3}{2}$$

$$x = 3$$

The period of the function is

$$\frac{2\pi}{C} = \frac{2\pi}{\left(\frac{\pi}{2}\right)} = 4$$

The phase shift of the function is

$$\frac{D}{C} = \frac{\left(\frac{\pi}{2}\right)}{\left(\frac{\pi}{2}\right)} = 1$$

Because the phase shift is positive, it is to the left.

The graph of the reciprocal function

$$y = 2\sin\left(\frac{\pi}{2}x + \frac{\pi}{2}\right)$$

is shown in Figure 3-29. Graphing the sine (or cosine) can make it easier to graph the cosecant (or secant).

Figure 3-29 Several periods of the cosecant function and the sine function.

$$y = 1/2 \, \csc(\pi x/2 + \pi/2)$$

Graphs of Special Trigonometric Functions

A pure tone, such as one produced by a tuning fork, is a wave form that looks like a sine curve. Sounds in general are more than just simple sine waves. They are combinations of sine waves and other functions. A vibrating string on a violin or fiddle is made up of a combination of several sine waves. The resulting wave form can be found by adding the ordinates of the respective sine waves. This can easily be seen if all the component wave forms are graphed on the same set of axes. Figures 3-30, 3-31, and 3-32 show the resulting wave form when two component wave forms are added together.

Example 25: Graph the functions $y = x$ and $y = 4 \sin x$ on the same coordinates and graph their sum.

Figure 3-30 Drawing for Example 25.

Example 26: Graph the functions $y = x/2$ and $y = 4 \sin x$ on the same coordinates and graph their sum.

Figure 3-31 Drawing for Example 26.

Example 27: Graph the functions $y = 10 \sin x$ and $y = 4 \sin 3x$ on the same coordinates and graph their sum.

Figure 3-32 Drawing for Example 27.

Chapter Checkout

Q&A

1. What is the radian measure of 83°?
2. What is the degree measure of 3.9 radians?
3. In radian measure, what is the size of the angle equal to one complete revolution around the unit circle?
4. What is the period of the sine function in degrees?
5. True or False: The sine and the tangent are negative in the fourth quadrant.
6. True or False: The sine and the tangent functions are odd functions.
7. True or False: The graph of the sine function and the graph of the cosine function are the same except for a shift to the left or right.
8. What is the period of the tangent function in radians?

Answers: 1. 1.45 radians **2.** 223.45° **3.** 2π **4.** 360° **5.** F **6.** T **7.** T **8.** π radians.

Chapter 4

TRIGONOMETRIC IDENTITIES

Chapter Checkin

❑ Defining several fundamental trigonometric identities

❑ Using sum and difference formulas to extend the fundamental identities

❑ Using double and half angle identities

❑ Understanding the tangent identities

❑ Using the product-to-sum and sum-to-product identities

Some equations are only true for one value of the unknown. Some equations are true for all values of the unknown. This second type of equation is called an identity because it is true for all values of the unknown variables. The knowledge of these identities is useful in solving more complex equations. As you explore new properties of trigonometric functions, new identities are established and then those can be used to establish still more identities, and so on.

Fundamental Identities

If an equation contains one or more variables and is valid for all replacement values of the variables for which both sides of the equation are defined, then the equation is known as an **identity**. The equation $x^2 + 2x = x(x + 2)$, for example, is an identity because it is valid for all replacement values of x.

If an equation is valid only for certain replacement values of the variable, then it is called a **conditional equation**. The equation $3x + 4 = 25$, for example, is a conditional equation because it is not valid for all replacement values of x. An equation that is said to be an identity without

stating any restrictions is, in reality, an identity only for those replacement values for which both sides of the identity are defined. For example, the identity

$$\frac{\sin\alpha}{\cos\alpha} = \tan\alpha$$

is valid only for those values of α for which both sides of the equation are defined.

The fundamental (basic) trigonometric identities can be divided into several groups. First are the **reciprocal identities**. These include

$$\cot\alpha = \frac{1}{\tan\alpha} \quad \sec\alpha = \frac{1}{\cos\alpha} \quad \csc\alpha = \frac{1}{\sin\alpha}$$

Next are the quotient identities. These include

$$\tan\alpha = \frac{\sin\alpha}{\cos\alpha} \quad \cot\alpha = \frac{\cos\alpha}{\sin\alpha}$$

Then there are the **cofunction identities**. These include

$$\sin\alpha = \cos(90° - \alpha) \qquad \cot\alpha = \tan(90° - \alpha)$$
$$\cos\alpha = \sin(90° - \alpha) \qquad \sec\alpha = \csc(90° - \alpha)$$
$$\tan\alpha = \cot(90° - \alpha) \qquad \csc\alpha = \sec(90° - \alpha)$$

Next there are the **identities for negatives**. These include

$$\sin(-\alpha) = -\sin\alpha$$
$$\cos(-\alpha) = \cos\alpha$$
$$\tan(-\alpha) = -\tan\alpha$$

Finally there are the **Pythagorean identities**. These include

$$\sin^2\alpha + \cos^2\alpha = 1$$
$$\tan^2\alpha + 1 = \sec^2\alpha$$
$$\cot^2\alpha + 1 = \csc^2\alpha$$

The second identity is obtained by dividing the first by $\cos^2\alpha$, and the third identity is obtained by dividing the first by $\sin^2\alpha$. The process of showing the validity of one identity based on previously known facts is called **proving the identity**. The validity of the foregoing identities follows directly from the definitions of the basic trigonometric functions and can be used to verify other identities.

No standard method for solving identities exists, but there are some general rules or strategies that can be followed to help guide the process:

1. Try to simplify the more complicated side of the identity until it is identical to the second side of the identity.

2. Try to transform both sides of an identity into an identical third expression.

3. Try to express both sides of the identity in terms of only sines and cosines; then try to make both sides identical.

4. Try to apply the Pythagorean identities as much as possible.

5. Try to use factoring and combining of terms, multiplying one side of the identity by an expression that is equal to 1, squaring both sides of the identity, and other algebraic techniques to manipulate equations.

Example 1: Use the basic trigonometric identities to determine the other five values of the trigonometric functions given that

$$\sin \alpha = \frac{7}{8} \quad and \quad \cos \alpha < 0$$

$$\sin^2 \alpha + \cos^2 \alpha = 1 \qquad\qquad \tan \alpha = \frac{\sin \alpha}{\cos \alpha}$$

$$\left(\frac{7}{8}\right)^2 + \cos^2 \alpha = 1$$

$$\cos^2 \alpha = 1 - \left(\frac{7}{8}\right)^2 \qquad\qquad \tan \alpha = \frac{\left(\frac{7}{8}\right)}{-\left(\frac{\sqrt{15}}{8}\right)}$$

$$\cos^2 \alpha = 1 - \frac{49}{64}$$

$$\cos^2 \alpha = \frac{15}{64} \qquad\qquad \tan \alpha = -\frac{7}{\sqrt{15}}$$

$$\cos \alpha = \sqrt{\frac{15}{64}} \qquad\qquad \tan \alpha = -\frac{7\sqrt{15}}{15}$$

$$\cos \alpha = -\frac{\sqrt{15}}{8}$$

$$\cot \alpha = -\frac{1}{\tan \alpha} \qquad \sec \alpha = -\frac{1}{\cos \alpha} \qquad \csc \alpha = -\frac{1}{\sin \alpha}$$

$$\cot \alpha = -\frac{1}{-\left(\frac{7\sqrt{15}}{15}\right)} \qquad \sec \alpha = \frac{1}{-\left(\frac{\sqrt{15}}{8}\right)} \qquad \csc \alpha = -\frac{1}{\frac{7}{8}}$$

$$\qquad\qquad\qquad\qquad\qquad\qquad\qquad\qquad\qquad\qquad \csc \alpha = \frac{8}{7}$$

$$\cot \alpha = -\frac{15}{7\sqrt{15}} \qquad \sec \alpha = \frac{8}{\sqrt{15}}$$

$$\cot \alpha = -\frac{\sqrt{15}}{7} \qquad \sec \alpha = \frac{8\sqrt{15}}{15}$$

Example 2: Verify the identity $\cos\alpha + \sin\alpha\tan\alpha = \sec\alpha$.

$$\cos\alpha + \sin\alpha\tan\alpha = \sec\alpha \qquad \text{identity to be verified}$$

$$(\cos\alpha) + \left(\frac{\cos\alpha}{\cos\alpha}\right) + (\sin\alpha)\left(\frac{\sin\alpha}{\cos\alpha}\right) = \sec\alpha \qquad \text{quotient identity}$$

$$\frac{\cos^2\alpha + \sin^2\alpha}{\cos\alpha} = \sec\alpha \qquad \text{algebraic manipulation}$$

$$\frac{1}{\cos\alpha} = \sec\alpha \qquad \text{Pythagorean identity}$$

$$\sec\alpha = \sec\alpha \qquad \text{reciprocal identity}$$

Example 3: Verify the identity

$$\frac{\tan\alpha}{\sec\alpha - 1} = \frac{\sec\alpha + 1}{\tan\alpha}$$

$$\frac{\tan\alpha}{\sec\alpha - 1} = \frac{\sec\alpha + 1}{\tan\alpha} \qquad \text{identity to be verified}$$

$$\frac{(\tan\alpha)(\sec\alpha + 1)}{(\sec\alpha - 1)(\sec\alpha + 1)} = \frac{(\sec\alpha + 1)}{\tan\alpha} \qquad \text{algebraic manipulation}$$

$$\frac{(\tan\alpha)(\sec\alpha + 1)}{\sec^2\alpha - 1} = \frac{\sec\alpha + 1}{\tan\alpha} \qquad \text{factoring rule}$$

$$\frac{(\tan\alpha)(\sec\alpha + 1)}{\tan^2\alpha} = \frac{\sec\alpha + 1}{\tan\alpha} \qquad \text{Pythagorean identity}$$

$$\frac{\sec\alpha + 1}{\tan\alpha} = \frac{\sec\alpha + 1}{\tan\alpha} \qquad \text{algebraic manipulation}$$

Example 4: Verify the identity

$$\frac{\sin^2\alpha + 2\cos\alpha - 1}{\sin^2\alpha + 3\cos\alpha - 3} = \frac{1}{1 - \sec\alpha}$$

$$\frac{\sin^2\alpha + 2\cos\alpha - 1}{\sin^2\alpha + 3\cos\alpha - 3} = \frac{1}{1 - \sec\alpha} \qquad \text{identity to be verified}$$

$$\frac{\left(1 - \cos^2\alpha\right) + 2\cos\alpha - 1}{\left(1 - \cos^2\alpha\right) + 3\cos\alpha - 3} = \frac{1}{1 - \sec\alpha} \qquad \text{Pythagorean identity}$$

$$\frac{-\left(\cos^2\alpha - 2\cos\alpha\right)}{-\left(\cos^2\alpha - 3\cos\alpha + 2\right)} = \frac{1}{1 - \sec\alpha} \qquad \text{combining term}$$

$$\frac{(\cos\alpha)(\cos\alpha - 2)}{(\cos\alpha - 1)(\cos\alpha - 2)} = \frac{1}{1 - \sec\alpha} \qquad \text{factoring}$$

$$\frac{\cos\alpha}{\cos\alpha - 1} = \frac{1}{1 - \sec\alpha} \qquad \text{algebraic manipulation}$$

$$\frac{\left(\dfrac{\cos\alpha}{\cos\alpha}\right)}{\left(\dfrac{\cos\alpha - 1}{\cos\alpha}\right)} = \frac{1}{1 - \sec\alpha} \qquad \text{algebraic manipulation}$$

$$\frac{1}{1 - \dfrac{1}{\cos\alpha}} = \frac{1}{1 - \sec\alpha} \qquad \text{algebraic manipulation}$$

$$\frac{1}{1 - \sec\alpha} = \frac{1}{1 - \sec\alpha} \qquad \text{reciprocal identity}$$

Addition Identities

The fundamental (basic) identities discussed in the previous section involved only one variable. The following identities, involving two variables, are called **trigonometric addition identities**.

$$\sin(\alpha + \beta) = \sin\alpha \cos\beta + \cos\alpha \sin\beta$$
$$\sin(\alpha - \beta) = \sin\alpha \cos\beta - \cos\alpha \sin\beta$$
$$\cos(\alpha + \beta) = \cos\alpha \cos\beta - \sin\alpha \sin\beta$$
$$\cos(\alpha - \beta) = \cos\alpha \cos\beta + \sin\alpha \sin\beta$$

These four identities are sometimes called the **sum identity for sine**, the **difference identity for sine**, the **sum identity for cosine**, and the **difference identity for cosine**, respectively. The verification of these four identities follows from the basic identities and the distance formula between points in the rectangular coordinate system. Explanations for each step of the proof will be given only for the first few examples that follow.

Example 5: Change $\sin 80° \cos 130° + \cos 80° \sin 130°$ into a trigonometric function in one variable (Figure 4-1).

$$\sin 80° \cos 130° + \cos 80° \sin 130°$$

$$= \sin(80° + 130°) \qquad \text{sum identity for sine}$$

$$= \sin(210°)$$

$$= -\sin(210° - 180°) \qquad \text{sine is negative in third quadrant}$$

$$-\sin 30°$$

$$= -\frac{1}{2}$$

Figure 4-1 Drawing for Example 5.

Additional identities can be derived from the sum and difference identities for cosine and sine.

Example 6: Verify that $\cos\left(180° - x\right) = -\cos x$

$$\cos\left(180° - x\right) = \cos 180° \cos x + \sin 180° \sin x$$

$$\cos\left(180° - x\right) = (-1)\cos x + (0)\sin x$$

$$\cos\left(180° - x\right) = -\cos x$$

Example 7: Verify that $\cos\left(180° + x\right) = -\cos x$

$$\cos\left(180° + x\right) = \cos 180° \cos x - \sin 180° \sin x$$

$$\cos\left(180° + x\right) = (-1)\cos x - (0)\sin x$$

$$\cos\left(180° + x\right) = -\cos x$$

Example 8: Verify that $\cos\left(360° - x\right) = \cos x$

$$\cos\left(360° - x\right) = \cos 360° \cos x + \sin 360° \sin x$$

$$\cos\left(360° - x\right) = (1)\cos x + (0)\sin x$$

$$\cos\left(360° - x\right) = \cos x$$

The preceding three examples verify three formulas known as the **reduction formulas for cosine**. These reduction formulas are useful in rewriting cosines of angles that are larger than 90° as functions of acute angles.

Example 9: Verify that $\sin\left(180° - x\right) = \sin x$

$$\sin\left(180° - x\right) = \sin 180° \cos x - \cos 180° \sin x$$

$$\sin\left(180° - x\right) = (0)\cos x - (-1)\sin x$$

$$\sin\left(180° - x\right) = \sin x$$

Example 10: Verify that $\sin\left(180° + x\right) = -\sin x$

$$\sin\left(180° + x\right) = \sin 180° \cos x + \cos 180° \sin x$$

$$\sin\left(180° + x\right) = (0)\cos x + (-1)\sin x$$

$$\sin\left(180° + x\right) = -\sin x$$

Example 11: Verify that $\sin\left(360° - x\right) = -\sin x$

$$\sin\left(360° - x\right) = \sin 360° \cos x - \cos 360° \sin x$$

$$\sin\left(360° - x\right) = (0)\cos x - (1)\sin x$$

$$\sin\left(360° - x\right) = -\sin x$$

The preceding three examples verify three formulas known as the **reduction formulas for sine**. These reduction formulas are useful in rewriting sines of angles that are larger than 90° as functions of acute angles.

To recap, the following are the reduction formulas (identities) for sine and cosine. They are valid for both degree and radian measure.

$$\cos\left(180° - x\right) = -\cos x \qquad \cos\left(\pi - x\right) = -\cos x$$

$$\cos\left(180° + x\right) = -\cos x \qquad \cos\left(\pi + x\right) = -\cos x$$

$$\cos\left(360° - x\right) = \cos x \qquad \cos\left(2\pi - x\right) = \cos x$$

$$\sin\left(180° - x\right) = \sin x \qquad \sin\left(\pi - x\right) = \sin x$$

$$\sin\left(180° + x\right) = -\sin x \qquad \sin\left(\pi + x\right) = -\sin x$$

$$\sin\left(360° - x\right) = -\sin x \qquad \sin\left(2\pi - x\right) = -\sin x$$

Example 12: Verify that $\sin 2x = 2 \sin x \cos x$.

$$\sin 2x = \sin(x+x)$$
$$\sin 2x = \sin x \cos x + \cos x \sin x$$
$$\sin 2x = 2 \sin x \cos x$$

Example 13: Write $\cos \beta \cos(\alpha - \beta) - \sin \beta \sin(\alpha - \beta)$ as a function of one variable.

$$\cos \beta \cos(\alpha - \beta) - \sin \beta \sin(\alpha - \beta)$$
$$= \cos\left[\beta + (\alpha - \beta)\right]$$
$$= \cos \alpha$$

Example 14: Write $\cos 303°$ in the form $\sin \beta$, where $0 < \beta < 90°$.

$$\cos 303° = \cos\left(360° - 303°\right)$$
$$= \cos 57°$$
$$= \sin\left(90° - 57°\right)$$
$$= \sin 33°$$

Example 15: Write $\sin 234°$ in the form $\cos 0 < \beta < 90°$.

$$\sin 234° = -\sin\left(360° - 234°\right)$$
$$= -\sin 126°$$
$$= -\sin\left(180° - 126°\right)$$
$$= -\sin 54°$$
$$= -\cos\left(90° - 54°\right)$$
$$= -\cos 36°$$

Example 16: Find $\sin(\alpha + \beta)$ if $\sin(\alpha + \beta)$ if $\sin \alpha = -\frac{4}{5}, \cos \beta = \frac{15}{17}$, and α and β are fourth quadrant angles.

First find $\cos \alpha$ and $\sin \beta$. The sine is negative and the cosine is positive in the fourth quadrant.

$$\sin^2\alpha + \cos^2\alpha = 1 \qquad \sin^2\beta + \cos^2\beta = 1$$

$$\left(\frac{-4}{5}\right)^2 + \cos^2\alpha = 1 \qquad \sin^2\beta + \left(\frac{15}{17}\right)^2 = 1$$

$$\cos^2 = 1 - \frac{16}{25} \qquad \sin^2\beta = 1 - \frac{225}{289}$$

$$\cos^2\alpha = \frac{9}{25} \qquad \sin^2\beta = \frac{64}{289}$$

$$\cos\alpha = \frac{3}{5} \qquad \sin\beta = -\frac{8}{17}$$

$$\sin(\alpha + \beta) = \sin\alpha\cos\beta + \cos\alpha\sin\beta$$

$$\sin(\alpha + \beta) = \left(-\frac{4}{5}\right)\left(\frac{15}{17}\right) + \left(\frac{3}{5}\right)\left(-\frac{8}{17}\right)$$

$$\sin(\alpha + \beta) = \left(-\frac{60}{85}\right) + \left(-\frac{24}{85}\right)$$

$$\sin(\alpha + \beta) = -\frac{84}{85}$$

Double-Angle and Half-Angle Identities

Special cases of the sum and difference formulas for sine and cosine yields what are known as the **double-angle identities** and the **half-angle identities**. First, using the sum identity for the sine,

$\sin 2\alpha = \sin(\alpha + \alpha)$

$\sin 2\alpha = \sin\alpha\cos\alpha + \cos\alpha\sin\alpha$

$\sin 2\alpha = 2\sin\alpha\cos\alpha$

Similarly for the cosine,

$$\cos 2\alpha = \cos(\alpha + \alpha)$$

$$\cos 2\alpha = \cos\alpha\cos\alpha - \sin\alpha\sin\alpha$$

$$\cos 2\alpha = \cos^2\alpha - \sin^2\alpha$$

Using the Pythagorean identity, $\sin^2\alpha + \cos^2\alpha = 1$, two additional cosine identities can be derived.

$$\cos 2\alpha = \cos^2\alpha - \sin^2\alpha$$

$$\cos 2\alpha = \left(-1\sin^2\alpha\right) - \sin^2\alpha$$

$$\cos 2\alpha = 1 - 2\sin^2\alpha$$

and

$$\cos 2\alpha = \cos^2 \alpha - \sin^2 \alpha$$
$$\cos 2\alpha = \cos^2 \alpha - \left(1 - \cos^2 \alpha\right)$$
$$\cos 2\alpha = 2\cos^2 \alpha - 1$$

The half-angle identities for the sine and cosine are derived from two of the cosine identities described earlier.

$$\cos 2\alpha = 2\cos^2 \alpha - 1$$
$$\cos 2\left(\frac{\beta}{2}\right) = 2\cos^2 \left(\frac{\beta}{2}\right) - 1$$
$$\cos \beta = 2\cos^2 \left(\frac{\beta}{2}\right) - 1$$
$$2\cos^2 \left(\frac{\beta}{2}\right) = 1 + \cos \beta$$

$$\cos 2\alpha = 1 - 2\sin^2 \alpha$$
$$\cos 2\left(\frac{\beta}{2}\right) = 1 - 2\sin^2 \left(\frac{\beta}{2}\right)$$
$$\cos \beta = 1 - 2\sin^2 \left(\frac{\beta}{2}\right)$$
$$2\sin^2 \left(\frac{\beta}{2}\right) = 1 - \cos \beta$$

$$\cos^2 \left(\frac{\beta}{2}\right) = \frac{1 + \cos \beta}{2} \qquad \sin^2 \left(\frac{\beta}{2}\right) = \frac{1 - \cos \beta}{2}$$
$$\cos \left(\frac{\beta}{2}\right) = \pm \sqrt{\frac{1 + \cos \beta}{2}} \qquad \sin \left(\frac{\beta}{2}\right) = \pm \sqrt{\frac{1 - \cos \beta}{2}}$$

The sign of the two preceding functions depends on the quadrant in which the resulting angle is located.

Example 17: Find the exact value for sin 105° using the half-angle identity.

In the following verification, remember that 105° is in the second quadrant, and sine functions in the second quadrant are positive. Also, 210° is in the third quadrant, and cosine functions in the third quadrant are negative. From Figure 4-2, the reference triangle of 210° in the third quadrant is a 30°-60°-90° triangle. Therefore, cos 210° = − cos 30°.

Figure 4-2 Drawing for Example 17.

Using the half-angle identity for sine,

$$\sin 105° = \sin \frac{210°}{2}$$

$$\sin 105° = \sqrt{\frac{1 - \cos 210°}{2}}$$

$$\sin 105° = \sqrt{\frac{1 - (-\cos 30°)}{2}}$$

$$\sin 105° = \sqrt{\frac{1 - \left(-\frac{\sqrt{3}}{2}\right)}{2}}$$

$$\sin 105° = \sqrt{\frac{2 + \sqrt{3}}{4}}$$

$$\sin 105° = \frac{\sqrt{2 + \sqrt{3}}}{2}$$

Example 18: Find the exact value for cos 165° using the half-angle identity.

In the following verification, remember that 165° is in the second quadrant, and cosine functions in the second quadrant are negative. Also, 330° is in the fourth quadrant, and cosine functions in the fourth quadrant are positive. From Figure 4-3, the reference triangle of 330° in the fourth quadrant is a 30°-60°-90° triangle. Therefore, cos 330° = cos 30°.

Figure 4-3 Drawing for Example 18.

Using the half-angle identity for the cosine,

$$\cos 165° = \cos \frac{330°}{2}$$

$$\cos 165° = \sqrt{\frac{1 + \cos 330°}{2}}$$

$$\cos 165° = \sqrt{\frac{1 + \left(\cos 30° \right)}{2}}$$

$$\cos 165° = -\sqrt{\frac{1 + \left[-\frac{\sqrt{3}}{2} \right]}{2}}$$

$$\cos 165° = -\sqrt{\frac{2 + \sqrt{3}}{4}}$$

$$\cos 165° = -\frac{\sqrt{2 + \sqrt{3}}}{2}$$

Example 19: Use the double-angle identity to find the exact value for $\cos 2x$ given that $\sin x = \sin x = \sqrt{5}/5$.

Because $\sin x$ is positive, angle x must be in the first or second quadrant. The sign of $\cos 2x$ will depend on the size of angle x. If $0° < x < 45°$ or $135° < x < 180°$, then $2x$ will be in the first or fourth quadrant and $\cos 2x$ will be positive. On the other hand, if $45° < x < 90°$ or $90° < x < 135°$, then $2x$ will be in the second or third quadrant and $\cos 2x$ will be negative.

$$\cos 2x = \pm \left(1 - 2\sin^2 x\right)$$

$$= \pm \left[1 - 2\left(\frac{\sqrt{5}}{5}\right)^2\right]$$

$$= \pm \left[1 - 2\left(\frac{5}{25}\right)\right]$$

$$= \pm \left(1 - \frac{2}{5}\right)$$

$$= \pm \frac{3}{5}$$

Example 20: Verify the identity $1 - \cos 2x = \tan x \sin 2x$.

$$1 - \cos 2x = \tan x \sin 2x$$

$$1 - \cos 2x = \left(\frac{\sin x}{\cos x}\right)\left(2\sin x \cos x\right)$$

$$1 - \cos 2x = 2\sin^2 x$$

$$1 - \left(1 - 2\sin^2 x\right) = 2\sin^2 x$$

$$2\sin^2 x = 2\sin^2 x$$

Tangent Identities

Formulas for the tangent function can be derived from similar formulas involving the sine and cosine. The sum identity for tangent is derived as follows:

$$\tan\left(\alpha + \beta\right) = \frac{\sin\left(\alpha + \beta\right)}{\cos\left(\alpha + \beta\right)}$$

$$\tan\left(\alpha + \beta\right) = \frac{\sin\alpha\sin\beta + \cos\alpha\sin\beta}{\cos\alpha\cos\beta - \sin\alpha\sin\beta}$$

$$\tan\left(\alpha + \beta\right) = \frac{\dfrac{\sin\alpha\cos\beta}{\cos\alpha\cos\beta} + \dfrac{\cos\alpha\sin\beta}{\cos\alpha\cos\beta}}{\dfrac{\cos\alpha\cos\beta}{\cos\alpha\cos\beta} - \dfrac{\sin\alpha\sin\beta}{\cos\alpha\cos\beta}}$$

$$\tan\left(\alpha + \beta\right) = \frac{\tan\alpha + \tan\beta}{1 - \tan\alpha\tan\beta}$$

To determine the **difference identity for tangent**, use the fact that $\tan(-\beta) = -\tan\beta$.

$$\tan(\alpha - \beta) = \tan\left[\alpha + (-\beta)\right]$$

$$\tan(\alpha - \beta) = \frac{\tan\alpha + \tan(-\beta)}{1 - \tan\alpha\tan(-\beta)}$$

$$\tan(\alpha - \beta) = \frac{\tan\alpha - \tan\beta}{1 + \tan\alpha\tan\beta}$$

Example 21: Find the exact value of $\tan 75°$.

Because $75° = 45° + 30°$

$$\tan 75° = \tan\left(45° + 30°\right)$$

$$\tan 75° = \frac{\tan 45° + \tan 30°}{1 - \tan 45°\tan 30°}$$

$$\tan 75° = \frac{1 + \dfrac{1}{\sqrt{3}}}{1 - (1)\left(\dfrac{1}{\sqrt{3}}\right)}$$

$$\tan 75° = \frac{\left(\dfrac{\sqrt{3}+1}{\sqrt{3}}\right)(\sqrt{3})}{\left(\dfrac{\sqrt{3}-1}{\sqrt{3}}\right)(\sqrt{3})}$$

$$\tan 75° = \frac{\sqrt{3}+1}{\sqrt{3}-1}$$

$$\tan 75° = \left(\frac{\sqrt{3}+1}{\sqrt{3}-1}\right)\left(\frac{\sqrt{3}+1}{\sqrt{3}+1}\right)$$

$$\tan 75° = \frac{3 + 2\sqrt{3} + 1}{3 - 1}$$

$$\tan 75° = 2 + \sqrt{3}$$

Example 22: Verify that tan $(180° - x) = -\tan x$.

$$\tan\left(180° - x\right) = \frac{\tan 180° - \tan x}{1 + \tan 180° \tan x}$$

$$\tan\left(180° - x\right) = \frac{0 - \tan x}{1 + (0) \tan x}$$

$$\tan\left(180° - x\right) = -\tan x$$

Example 23: Verify that tan $(180° + x) = \tan x$.

$$\tan\left(180° + x\right) = \frac{\tan 180° + \tan x}{1 - \tan 180° \tan x}$$

$$\tan\left(180° + x\right) = \frac{0 + \tan x}{1 - (0) \tan x}$$

$$\tan\left(180° + x\right) = \tan x$$

Example 24: Verify that tan $(360° - x) = -\tan x$.

$$\tan\left(360° - x\right) = \frac{\tan 360° + \tan x}{1 + \tan 360° \tan x}$$

$$\tan\left(360° - x\right) = \frac{0 - \tan x}{1 + (0) \tan x}$$

$$\tan\left(360° - x\right) = -\tan x$$

The preceding three examples verify three formulas known as the **reduction identities for tangent**. These reduction formulas are useful in rewriting tangents of angles that are larger than 90° as functions of acute angles.

The **double-angle identity for tangent** is obtained by using the sum identity for tangent.

$$\tan\left(2\alpha\right) = \tan\left(\alpha + \alpha\right)$$

$$\tan\left(2\alpha\right) = \frac{\tan\alpha + \tan\alpha}{1 - \tan\alpha \tan\alpha}$$

$$\tan\left(2\alpha\right) = \frac{2\tan\alpha}{1 - \tan^2\alpha}$$

The **half-angle identity for tangent** can be written in three different forms.

$$\tan\frac{\alpha}{2} = \pm\sqrt{\frac{1 - \cos\alpha}{1 + \cos\alpha}}$$

$$\tan\frac{\alpha}{2} = \frac{\sin\alpha}{1 + \cos\alpha}$$

$$\tan\frac{\alpha}{2} = \frac{1 - \cos\alpha}{\sin\alpha}$$

In the first form, the sign is determined by the quadrant in which the angle $\alpha/2$ is located.

Example 25: Verify the identity

$$\tan\frac{\alpha}{2} = \pm\sqrt{\frac{1-\cos\alpha}{1+\cos\alpha}}$$

$$\tan\frac{\alpha}{2} = \frac{\sin\frac{\alpha}{2}}{\cos\frac{\alpha}{2}}$$

$$\tan\frac{\alpha}{2} = \frac{\pm\sqrt{\dfrac{1-\cos\alpha}{2}}}{\pm\sqrt{\dfrac{1+\cos\alpha}{2}}}$$

$$\tan\frac{\alpha}{2} = \frac{\pm\dfrac{\sqrt{1-\cos\alpha}}{\sqrt{2}}}{\pm\dfrac{\sqrt{1+\cos\alpha}}{\sqrt{2}}}$$

$$\tan\frac{\alpha}{2} = \pm\sqrt{\frac{1-\cos\alpha}{1+\cos\alpha}}$$

Example 26: Verify the identity $\tan(\alpha/2) = (1 - \cos\alpha)/\sin\alpha$.

$$\tan\frac{\alpha}{2} = \frac{\sin\frac{\alpha}{2}}{\cos\frac{\alpha}{2}}$$

$$\tan\frac{\alpha}{2} = \left(\frac{\sin\frac{\alpha}{2}}{\cos\frac{\alpha}{2}}\right)\left(\frac{2\sin\frac{\alpha}{2}}{2\cos\frac{\alpha}{2}}\right)$$

$$\tan\frac{\alpha}{2} = \frac{2\sin^2\frac{\alpha}{2}}{2\sin\frac{\alpha}{2} = \cos\frac{\alpha}{2}}$$

$$\tan\frac{\alpha}{2} = \frac{2\left(\dfrac{1-\cos\alpha}{2}\right)}{\sin 2\frac{\alpha}{2}}$$

$$\tan\frac{\alpha}{2} = \frac{1-\cos\alpha}{\sin\alpha}$$

Example 27: Verify the identity $\tan(\alpha/2) = \sin\alpha/(1 + \cos\alpha)$.

Begin with the identity in Example 26.

$$\tan\frac{\alpha}{2} = \frac{1 - \cos\alpha}{\sin\alpha}$$

$$\tan\frac{\alpha}{2} = \left(\frac{1 - \cos\alpha}{\sin\alpha}\right)\left(\frac{1 + \cos\alpha}{1 + \cos\alpha}\right)$$

$$\tan\frac{\alpha}{2} = \frac{1 - \cos^2\alpha}{(\sin\alpha)(1 + \cos\alpha)}$$

$$\tan\frac{\alpha}{2} = \frac{\sin^2\alpha}{(\sin\alpha)(1 + \cos\alpha)}$$

$$\tan\frac{\alpha}{2} = \frac{\sin\alpha}{1 + \cos\alpha}$$

Example 28: Use a half-angle identity for the tangent to find the exact value for tan 15°.

What follows are two alternative solutions.

Solution A

$$\tan 15° = \tan\frac{30°}{2}$$

$$\tan 15° = \frac{1 - \cos 30°}{\sin 30°}$$

$$\tan 15° = \frac{\left(1 - \frac{\sqrt{3}}{2}\right)}{\left(\frac{1}{2}\right)}$$

Solution B

$$\tan 15° = \tan\frac{30°}{2}$$

$$\tan 15° = \frac{\sin 30°}{1 + \cos 30°}$$

$$\tan 15° = \frac{\left(\frac{1}{2}\right)}{\left(1 + \frac{\sqrt{3}}{2}\right)}$$

Solution A

$$\tan 15° = \left(\frac{2 + \sqrt{3}}{2}\right)\left(\frac{2}{1}\right)$$

$$\tan 15° = 2 - \sqrt{3}$$

Solution B

$$\tan 15° = \left(\frac{1}{2}\right)\left(\frac{2}{2 + \sqrt{3}}\right)$$

$$\tan 15° = \frac{1}{2 + \sqrt{3}}$$

$$\tan 15° = \left(\frac{1}{2 + \sqrt{3}}\right)\left(\frac{2 - \sqrt{3}}{2 - \sqrt{3}}\right)$$

$$\tan 15° = 2 - \sqrt{3}$$

Product–Sum and Sum–Product Identities

The process of converting sums into products or products into sums can make a difference between an easy solution to a problem and no solution

at all. Two sets of identities can be derived from the sum and difference identities that help in this conversion. The following set of identities is known as the **product-sum identities**.

$$\sin\alpha\cos\beta = \frac{1}{2}\Big[\sin(\alpha+\beta) + \sin(\alpha-\beta)\Big]$$

$$\cos\alpha\sin\beta = \frac{1}{2}\Big[\sin(\alpha+\beta) - \sin(\alpha-\beta)\Big]$$

$$\sin\alpha\sin\beta = \frac{1}{2}\Big[\cos(\alpha-\beta) - \cos(\alpha+\beta)\Big]$$

$$\cos\alpha\cos\beta = \frac{1}{2}\Big[\cos(\alpha+\beta) + \cos(\alpha-\beta)\Big]$$

These identities are valid for degree or radian measure whenever both sides of the identity are defined.

Example 29: Verify that $\sin\alpha\cos\beta =$
$\sin\alpha\cos\beta = \frac{1}{2}\Big[\sin(\alpha+\beta) + \sin(\alpha-\beta)\Big]$.

Start by adding the sum and difference identities for the sine.

$$\sin(\alpha+\beta) = \sin\alpha\cos\beta + \cos\alpha\sin\beta$$

$$\sin(\alpha-\beta) = \sin\alpha\cos\beta - \cos\alpha\sin\beta$$

$$\sin(\alpha+\beta) + \sin(\alpha-\beta) = 2\sin\alpha\cos\beta$$

$$\sin\alpha\cos\beta = \frac{1}{2}\Big[\sin(\alpha+\beta) + \sin(\alpha-\beta)\Big]$$

The other three product-sum identities can be verified by adding or subtracting other sum and difference identities.

Example 30: Write $\cos 3x \cos 2x$ as a sum.

$$\cos\alpha\cos\beta = \frac{1}{2}\Big[\cos(\alpha+\beta) + \cos(\alpha-\beta)\Big]$$

$$\cos 3x\cos 2x = \frac{1}{2}\Big[\cos(3x+2x) + \cos(3x-2x)\Big]$$

$$\cos 3x\cos 2x = \frac{1}{2}(\cos 5x + \cos x)$$

$$\cos 3x\cos 2x = \frac{\cos 5x}{2} + \frac{\cos x}{2}$$

Alternate forms of the product-sum identities are the **sum-product identities**.

$$\sin\alpha + \sin\beta = 2\sin\frac{\alpha+\beta}{2}\cos\frac{\alpha-\beta}{2}$$

$$\sin\alpha - \sin\beta = 2\cos\frac{\alpha+\beta}{2}\sin\frac{\alpha-\beta}{2}$$

$$\cos\alpha + \cos\beta = 2\cos\frac{\alpha+\beta}{2}\cos\frac{\alpha-\beta}{2}$$

$$\cos\alpha - \cos\beta = -2\sin\frac{\alpha+\beta}{2}\sin\frac{\alpha-\beta}{2}$$

These identities are valid for degree or radian measure whenever both sides of the identity are defined.

Example 31: Verify that $\sin x + \sin y = 2\sin\frac{x+y}{2}\cos\frac{x-y}{2}$.

$$\sin\alpha\,\cos\beta = \frac{1}{2}\Big[\sin(\alpha+\beta) + \sin(\alpha-\beta)\Big]$$

Solve for α by adding the following two equations and then dividing by 2. Solve for β by subtracting the two equations and then dividing by 2.

If $x = \alpha + \beta$ and $y = \alpha - \beta$

Then $\alpha = \frac{x+y}{2}$ and $\beta = \frac{x-y}{2}$

$$\sin\frac{x+y}{2}\cos\frac{x-y}{2} = \frac{1}{2}\big(\sin x + \sin y\big)$$

$$\sin x + \sin y = 2\sin\frac{x+y}{2}\cos\frac{x-y}{2}$$

Example 32: Write the difference $\cos 8\alpha - \cos 2\alpha$ as a product.

$$\cos\alpha - \cos\beta = -2\sin\frac{\alpha+\beta}{2}\sin\frac{\alpha-\beta}{2}$$

$$\cos 8\alpha - \cos 2\alpha = -2\sin 5\alpha\sin 3\alpha$$

Example 33: Find the exact value of $\sin 75° + \sin 15°$.

$$\sin x + \sin y = 2\sin\frac{x+y}{2}\cos\frac{x-y}{2}$$

$$\sin 75° + \sin 15° = 2\sin 45°\cos 30°$$

$$\sin 75° + \sin 15° = 2\left(\frac{1}{\sqrt{2}}\right)\left(\frac{\sqrt{3}}{2}\right)$$

$$\sin 75° + \sin 15° = \frac{\sqrt{3}}{\sqrt{2}}$$

$$\sin 75° + \sin 15° = \frac{\sqrt{6}}{2}$$

Chapter Checkout

Q&A

1. Use the fundamental trig identities to find the remaining five trig functions if $\cos \alpha = 2/3$ and $\tan \alpha < 0$.

2. Establish this identity: $\sec \theta - \cos \theta = \tan \theta \sin \theta$.

3. True or False: The sum and difference formulas can be used to find the exact value of the cosine of $75°$.

4. True or False: When establishing an identity, you may only work on one side of the identity at a time.

5. Establish this identity: $\csc \theta = -\dfrac{3\sqrt{5}}{5}$.

Answers: 1. $\sin \theta = -\dfrac{\sqrt{5}}{3}$, $\tan \theta = -\dfrac{\sqrt{5}}{2}$, $\cot \theta = -\dfrac{2\sqrt{5}}{5}$,

$$\sec \theta = \frac{3}{2}, \ \csc \theta = -\frac{3\sqrt{5}}{5}$$

2. $\sec \theta - \cos \theta = \tan \theta \sin \theta$

$$\frac{1}{\cos \theta} - \frac{\cos^2 \theta}{\cos \theta} =$$

$$\frac{1 - \cos^2 \theta}{\cos \theta} =$$

$$\frac{\sin^2 \theta}{\cos \theta} =$$

$$\frac{\sin \theta}{\cos \theta} \sin \theta =$$

$$\tan \theta \sin \theta =$$

3. T **4.** T

5. $\dfrac{\cos \theta + \sin \theta}{\sin \theta} = 1 + \cot \theta$

$$\frac{\cos \theta}{\sin \theta} + \frac{\sin \theta}{\sin \theta} =$$

$$\cot \theta + 1 =$$

$$1 + \cot \theta$$

Chapter 5
VECTORS

Chapter Checkin

❏ Understanding vectors and their components

❏ Applying the tip tail rule for combining vectors

❏ Using vectors to solve problems

❏ Defining vector operations involving addition and the dot product rule

In the physical world, some quantities, such as mass, length, age, and value, can be represented by only magnitude. Other quantities, such as speed and force, also involve direction. You can use vectors to represent those quantities that involve both magnitude and direction. One common use of vectors involves finding the actual speed and direction of an aircraft given its air speed and direction and the speed and direction of a tailwind. Another common use of vectors involves finding the resulting force on an object being acted upon by several separate forces.

Vector Operations

Any quantity that has both size and direction is called a **vector quantity**. If A and B are two points that are located in a plane, the **directed line segment** from point A to point B is denoted by \overrightarrow{AB}. Point A is the **initial point**, and point B is the **terminal point**.

A **geometric vector** is a quantity that can be represented by a directional line segment. From this point on, a vector will be denoted by a boldface letter, such as **v** or **u**. The **magnitude** of a vector is the length of the directed line segment. The magnitude is sometimes called the **norm**. Two vectors have the *same direction* if they are parallel and point in the same direction. Two vectors have *opposite directions* if they are parallel and point

in opposite directions. A vector that has no magnitude and points in any direction is called the **zero vector**. Two vectors are said to be **equivalent vectors** if they have the same magnitude and same direction.

Figure 5-1 demonstrates **vector addition** using the tail-tip rule. To add vectors **v** and **u**, translate vector **u** so that the initial point of **u** is at the terminal point of **v**. The resulting vector from the initial point of **v** to the terminal point of **u** is the vector **v + u** and is called the **resultant**. The vectors **v** and **u** are called the **components** of the vector **v + u**. If the two vectors to be added are not parallel, then the **parallelogram rule** can also be used. In this case, the initial points of the vectors are the same, and the resultant is the diagonal of the parallelogram formed by using the two vectors as adjacent sides of the parallelogram.

Figure 5-1 Example of vector addition.

In order to multiply a vector **u** by a real number q, multiply the length of **u** by $|q|$ and reverse the direction of **u** if $q < 0$. This is called **scalar multiplication**. If a vector **u** is multiplied by -1, the resulting vector is designated as $-\mathbf{u}$. It has the same magnitude as **u** but opposite direction. Figure 5-2 demonstrates the use of scalars.

Figure 5-2 Examples of vectors.

Example 1: A plane is traveling due west with an air speed of 400 miles per hour. There is a tailwind blowing in a southwest direction at 50 miles per hour. Draw a diagram that represents the plane's ground speed and direction (Figure 5-3).

Figure 5-3 Drawing for Example 1– vector representation.

The vector represented in the preceding example is known as a **velocity vector**. The **bearing** of a vector **v** is the angle measured clockwise from due north to **v**. In the example, the bearing of the plane is 270° and the bearing of the wind is 225°. Redrawing the figure as a triangle using the tail-tip rule, the length (ground speed of the plane) and bearing of the resultant can be calculated (Figure 5-4).

Figure 5-4 Drawing for Example 1– angle representation.

First, use the law of cosines to find the magnitude of the resultant.

$$c^2 = a^2 + b^2 - 2ab\cos C$$
$$c^2 = (400)^2 + (50)^2 - 2(400)(50)\cos 135°$$
$$c^2 \approx 160,000 + 2,500 - (40,000)(-.7071)$$
$$c^2 \approx 162,500 - 28,284$$
$$c^2 \approx 190,784$$
$$c \approx 436.8$$

Then, use the law of sines to find the bearing.

$$\frac{\sin C}{c} = \frac{\sin B}{b}$$

$$\frac{\sin 135°}{436.8} \approx \frac{\sin B}{50}$$

$$\sin B \approx \frac{(0.7071)(50)}{436.8}$$

$$\sin B \approx 0.0809$$

$$B \approx 4.64°$$

The bearing, β, is therefore 270° – 4.64°, or approximately 265.4°.

Example 2: A plane flies at 300 miles per hour. There is a wind blowing out of the southeast at 86 miles per hour with a bearing of 320°. At what bearing must the plane head in order to have a true bearing (relative to the ground) of 14°? What will be the plane's groundspeed (Figure 5-5)?

Figure 5-5 Drawing for Example 2.

Use the law of sines to calculate the bearing and the groundspeed. Because these alternate interior angles are congruent, the 54° angle is the sum of the 14° angle and the 40° angle.

$$\frac{\sin \alpha}{86} = \frac{\sin 54°}{300}$$

$$\sin \alpha \approx \frac{(86)(0.809)}{300}$$

$$\sin \alpha \approx 0.232$$

$$\alpha \approx 13.4°$$

$$\frac{\sin 54°}{300} = \frac{\sin(180 - 54 - 13.4)°}{x}$$

$$x \approx \frac{(300)\left(\sin 112.6°\right)}{\sin 54°}$$

$$x \approx \frac{(300)(0.923)}{0.809}$$

$$x \approx 342.3$$

Therefore, the bearing of the plane should be 14° + 13.4° = 27.4°. The groundspeed of the plane is 342.3 miles per hour.

Any vector can be broken down into two **component vectors**, a horizontal component and a vertical component. These component vectors are called **projections** (Figure 5-6).

Figure 5-6 Example of projections.

$$\cos 71^\circ = \frac{|x|}{26} \qquad \sin 71^\circ = \frac{|y|}{26}$$

$$|x| = 26\cos 71^\circ \qquad |y| = 26\sin 71^\circ$$

$$|x| \approx 8.46 \qquad |y| = 24.6$$

Example 3: A force of 11 pounds and a force of 6 pounds act on an object at an angle of 41° with respect to one another. What is the magnitude of the resultant force, and what angle does the resultant force form with the 11-pound force (Figure 5-7)?

Figure 5-7 Drawing for Example 3.

First, use the Law of Cosines to find the magnitude of the resultant force.

$$c^2 = a^2 + b^2 - 2ab\cos C$$

$$|v|^2 = 6^2 + 11^2 - (2)(6)(11)\left(\cos 139^\circ\right)$$

$$|v|^2 \approx 36 + 121 - (132)(-0.755)$$

$$|v|^2 \approx 256.7$$

$$|v| \approx 16.02$$

Next, use the Law of Sines.

$$\frac{\sin A}{a} = \frac{\sin B}{b}$$

$$\frac{\sin \alpha}{6} = \frac{\sin 139°}{16.02}$$

$$\sin \alpha \approx \frac{(6)0.656}{16.02}$$

$$\sin \alpha \approx 0.246$$

$$\alpha \approx 14.24°$$

Thus the resultant force is 16.02 pounds, and this force makes an angle of 14.24° with the 11-pound force.

Vectors in the Rectangular Coordinate System

The following discussion is limited to vectors in a two-dimensional coordinate plane, although the concepts can be extended to higher dimensions.

If vector \overrightarrow{AB} is shifted so that its initial point is at the origin of the rectangular coordinate plane, it is said to be in **standard position**. If vector \overrightarrow{OP} is equal to vector \overrightarrow{AB} and has its initial point at the origin, it is said to be the standard vector for \overrightarrow{AB}. Other names for the standard vector include radius vector and position vector (Figure 5-8).

Figure 5-8 Vectors drawn on a plane.

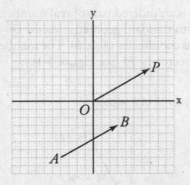

Vector \overrightarrow{OP} is the standard vector for all vectors in the plane with the same direction and magnitude as \overrightarrow{OP}. In order to find the standard vector for a

a geometric vector in the coordinate plane, only the coordinates of point *P* must be found because point *0* is at the origin. If the coordinates of point *A* are (x_a, y_a) and the coordinates of point *B* are (x_b, y_b), then the coordinates of point *P* are $(x_b - x_a, y_b - y_a)$.

Example 4: If the endpoints of a vector \overrightarrow{AB} have coordinates of *A* (−2, −7) and B (3, 2), then what are the coordinates of point *P* such that \overrightarrow{OP} is a standard vector and $\overrightarrow{OP} = \overrightarrow{AB}$ (see Figure 5-9)?

Figure 5-9 Drawing for Example 4.

If the coordinates of point *P* are (x, y),

$$x = x_b - x_a = 3 - (-2) = 5$$
$$y = y_b - y_a = 2 - (-7) = 9$$

An **algebraic vector** is an ordered pair of real numbers. An algebraic vector that corresponds to standard geometric vector \overrightarrow{OP} is denoted as $\langle a, b \rangle$ if terminal point P has coordinates of *(a, b)*. The numbers *a* and *b* are called the **components** of vector $\langle a, b \rangle$ (see Figure 5-10).

Figure 5-10 Components of a vector.

If a, b, c, and d are all real numbers such that $a = c$ and $b = d$, then vector $\mathbf{v} = \langle a, b \rangle$ and vector $\mathbf{u} = \langle c, d \rangle$ are said to be equal. That is, algebraic vectors with equal corresponding components are equal. If both components of a vector are equal to zero, the vector is said to be the **zero vector**. The **magnitude** of a vector $\mathbf{v} = \langle a, b \rangle$ is $|v| = \sqrt{a^2 + b^2}$.

Example 5: What is the magnitude of vector $\mathbf{u} = \langle 3, -5 \rangle$?

$$|u| = \sqrt{a^2 + b^2}$$
$$|u| = \sqrt{3^2 + (-5)^2}$$
$$|u| = \sqrt{9 + 25}$$
$$|u| = \sqrt{34}$$

Vector addition is defined as adding corresponding components of vectors—that is, if $\mathbf{v} = \langle a, b \rangle$ and $\mathbf{u} = \langle c, d \rangle$, then $\mathbf{v} + \mathbf{u} = \langle a + c, b + d \rangle$ (Figure 5-11).

Figure 5-11 Vector addition.

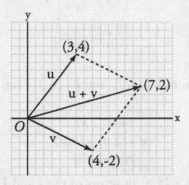

Scalar multiplication is defined as multiplying each component by a constant—that is, if $\mathbf{v} = \langle a, b \rangle$ and q is a constant, then $q\mathbf{v} = q \langle a, b \rangle = \langle qa, qb \rangle$

Example 6: If $\mathbf{v} = \langle 8, -2 \rangle$ and $\mathbf{w} = \langle 3, 7 \rangle$ then find $5\mathbf{v} - 2\mathbf{w}$.

$$5v - 2\mathbf{w} = \langle 8, -2 \rangle - 2\langle 3, 7 \rangle$$
$$5v - 2\mathbf{w} = \langle 40, -10 \rangle - \langle 6, 14 \rangle$$
$$5v - 2\mathbf{w} = \langle 34, -24 \rangle$$

A **unit vector** is a vector whose magnitude is 1. A unit vector **v** with the same direction as a nonzero vector **u** can be found as follows:

$$v = \frac{1}{|u|} u$$

Example 7: Find a unit vector **v** with the same direction as the vector **u** given that $\mathbf{u} = \langle 7, -1 \rangle$.

$$|u| = \sqrt{7^2 + (-1)^2}$$
$$|u| = \sqrt{49 + 1}$$
$$|u| = \sqrt{50}$$
$$|u| = 5\sqrt{2}$$
$$v = \frac{1}{|u|} u$$
$$v = \frac{1}{5\sqrt{2}} \langle 7, -1 \rangle$$
$$v = \left\langle \frac{7}{5\sqrt{2}}, \frac{-1}{5\sqrt{2}} \right\rangle$$
$$v = \left\langle \frac{7\sqrt{2}}{10}, \frac{-\sqrt{2}}{10} \right\rangle$$

Two special unit vectors, $\mathbf{i} = \langle 1, 0 \rangle$ and $\mathbf{j} = \langle 0, 1 \rangle$, can be used to express any vector $\mathbf{v} = \langle a, b \rangle$.

$$\mathbf{v} = \langle a, b \rangle$$
$$\mathbf{v} = \langle a, 0 \rangle + \langle 0, b \rangle$$
$$\mathbf{v} = a\mathbf{i} + b\mathbf{j}$$

Example 8: Write $\mathbf{u} = \langle 5, 3 \rangle$ in terms of the **i** and **j** unit vectors (Figure 5-12).

$$\mathbf{u} = \langle 5, 3 \rangle$$
$$\mathbf{u} = 5\langle 1, 0 \rangle + 3\langle 0, 1 \rangle$$
$$\mathbf{u} = 5\mathbf{i} + 3\mathbf{j}$$

Figure 5-12 Drawing for Example 8.

Vectors exhibit algebraic properties similar to those of real numbers (Table 5-1).

Table 5-1 Properties of Vectors

Associative property	$\mathbf{u} + (\mathbf{v} + \mathbf{w}) = (\mathbf{u} + \mathbf{v}) + \mathbf{w}$	$a(b\mathbf{v}) = (ab)\mathbf{v}$
Commutative property	$\mathbf{u} + \mathbf{v} = \mathbf{v} + \mathbf{u}$	
Distributive property	$a(\mathbf{u} + \mathbf{v}) = a\mathbf{u} + a\mathbf{v}$	$(a + b)\mathbf{v} = a\mathbf{v} + b\mathbf{v}$
Identity	$\mathbf{v} + \mathbf{0} = \mathbf{0} + \mathbf{v} = \mathbf{v}$	$1\mathbf{v} = \mathbf{v}$
Inverse	$\mathbf{v} + (-\mathbf{v}) = \mathbf{0}$	

Example 9: Find $4\mathbf{u} + 5\mathbf{v}$ if $\mathbf{u} = 7\mathbf{i} - 3\mathbf{j}$ and $\mathbf{v} = -2\mathbf{i} + 5\mathbf{j}$.

$$4\mathbf{u} + 5\mathbf{v} = 4(7\mathbf{i} - 3\mathbf{j}) + 5(-2\mathbf{i} + 5\mathbf{j})$$
$$4\mathbf{u} + 5\mathbf{v} = 28\mathbf{i} - 12\mathbf{j} - 10\mathbf{i} + 25\mathbf{j}$$
$$4\mathbf{u} + 5\mathbf{v} = (28 - 10)\mathbf{i} + (-12 + 25)\mathbf{j}$$
$$4\mathbf{u} + 5\mathbf{v} = 18\mathbf{i} + 13\mathbf{j}$$

Given two vectors, $\mathbf{u} = \langle a, b \rangle = a\mathbf{i} + b\mathbf{j}$ and $\mathbf{v} = \langle c, d \rangle = c\mathbf{i} + d\mathbf{j}$, the **dot product**, written as $\mathbf{u} \cdot \mathbf{v}$, is the scalar quantity $\mathbf{u} \cdot \mathbf{v} = ac + bd$. If \mathbf{u}, \mathbf{v}, and \mathbf{w} are vectors and q is a real number, then dot products exhibit the following properties:

$$\mathbf{u} \cdot \mathbf{v} = \mathbf{v} \cdot \mathbf{u}$$

$$\mathbf{u} \cdot (\mathbf{v} + \mathbf{w}) = \mathbf{u} \cdot \mathbf{v} + \mathbf{u} \cdot \mathbf{w}$$

$$(\mathbf{u} + \mathbf{v}) \cdot \mathbf{w} = \mathbf{u} \cdot \mathbf{w} + \mathbf{v} \cdot \mathbf{w}$$

$$q(\mathbf{u} \cdot \mathbf{v}) = (q\mathbf{u}) \cdot \mathbf{v} = \mathbf{u} \cdot (q\mathbf{v})$$

$$\mathbf{u} \cdot 0 = 0$$

$$\mathbf{v} \cdot \mathbf{v} = |\mathbf{v}|^2$$

$$\mathbf{u} \cdot \mathbf{v} = |\mathbf{u}||\mathbf{v}|\cos\alpha$$

The last property, $\mathbf{u} \cdot \mathbf{v} = |\mathbf{u}||\mathbf{v}|\cos\alpha$, can be used to find the angle between the two nonzero vectors \mathbf{u} and \mathbf{v}. If two vectors are perpendicular to each other and form a 90° angle, they are said to be **orthogonal**. Because cos 90°= 0, the dot product of any two orthogonal vectors is 0.

Example 10: Given that $\mathbf{u} = \langle 5, -3 \rangle$ and $\mathbf{v} = \langle 6, 10 \rangle$, show that \mathbf{u} and \mathbf{v} are orthogonal by demonstrating that the dot product of \mathbf{u} and \mathbf{v} is equal to zero.

$$\mathbf{u} \cdot \mathbf{v} = \langle 5, -3 \rangle \cdot \langle 6, 10 \rangle$$

$$\mathbf{u} \cdot \mathbf{v} = 30 + (-30)$$

$$\mathbf{u} \cdot \mathbf{v} = 0$$

Example 11: What is the angle between $\mathbf{u} = \langle 5, -2 \rangle$ and $\mathbf{v} = \langle 6, 11 \rangle$?

$$\cos\alpha = \frac{\mathbf{u} \cdot \mathbf{v}}{|\mathbf{u}||\mathbf{v}|}$$

$$\cos\alpha = \frac{\langle 5, -2 \rangle \cdot \langle 6, 11 \rangle}{\sqrt{5^2 + (-2)^2}\sqrt{6^2 + 11^2}}$$

$$\cos\alpha = \frac{8}{\sqrt{29}\sqrt{157}}$$

$$\cos\alpha \approx 0.119$$

$$\alpha \approx 83.2°$$

An object is said to be in a state of **static equilibrium** if all the force vectors acting on the object add up to zero.

Example 12: A tightrope walker weighing 150 pounds is standing closer to one end of the rope than the other. The shorter length of rope deflects 5° from the horizontal. The longer length of rope deflects 3°. What is the tension on each part of the rope?

Draw a force diagram with all three force vectors in standard position (Figure 5-13).

Figure 5-13 Drawing for Example 12.

$$u = \left(-|u|\cos5°\right)i + \left(|u|\sin5°\right)j$$
$$v = \left(|v|\cos3°\right)i + \left(|v|\sin3°\right)j$$
$$w = -150j$$

The sum of the force vectors must be zero for each component.

For the **i** component: $-|u|\cos5° + |v|\cos3° = 0$

For the **j** component: $|u|\sin5° + |v|\cos3° - 150 = 0$

Solve these two equations for $|u|$ and $|v|$:

$$-|u|\cos5° + |v|\cos3° = 0$$
$$|u|\sin5° + |v|\cos3° - 150 = 0$$

Substituting the values for the sines and cosines:

$$-0.9962|u| + 0.9986|v| = 0$$
$$0.0872|u| + 0.0523|v| - 150 = 0$$

Multiply the first equation by 0.0872 and the second by 0.9962:

$$-0.0869|\mathbf{u}| + 0.0871|\mathbf{v}| = 0$$

$$0.0869|\mathbf{u}| + 0.0521|\mathbf{v}| = 149.43$$

Add the two equations and solve for $|\mathbf{v}|$:

$$0.13929|\mathbf{v}| = 149.43$$

$$|\mathbf{v}| = 1073\,\text{lbs}$$

Substitute and solve for $|\mathbf{u}|$:

$$|\mathbf{u}| = 1076\,\text{lbs}$$

Chapter Checkout

Q&A

1. An airplane is traveling with airspeed of 225 mph at a bearing of 205°. A 60 mph wind is blowing with a bearing of 100°. What is the resultant ground speed and direction of the plane?

2. A force of 22 pounds and a force of 35 pounds act on an object at an angle of 32° with respect to one another. What is the resultant force on the object?

3. If the endpoints of a vector \overrightarrow{AB} have coordinates of A(-4, 6), and B(10,4), then what are the coordinates of a point P such that \overrightarrow{OP} is a standard vector and $\overrightarrow{OP} = \overrightarrow{AB}$?

4. True or False: The dot product of two orthogonal vectors is always zero.

5. What is the angle between vector \mathbf{u} = <5, 7> and \mathbf{v} = <-6, 6> ?

Answers: 1. 217 mph, bearing 189.53° **2.** 54.91 pounds **3.** (14,−2) **4.** True **5.** 80.54°.

Chapter 6

POLAR COORDINATES AND COMPLEX NUMBERS

Chapter Checkin

❑ Defining polar coordinates in terms of rectangular coordinates

❑ Converting between polar and rectangular coordinates

❑ Recognizing the graphs of some important polar curves

❑ Defining complex numbers in terms of polar coordinates

❑ Converting between complex numbers and polar coordinates

❑ Defining and using De Moivre's Theorem

Many systems and styles of measure are in common use today. When graphing on a flat surface, the rectangular coordinate system and the polar coordinate system are the two most popular methods for drawing the graphs of relations. Polar coordinates are best used when periodic functions are considered. Although either system can usually be used, polar coordinates are especially useful under certain conditions.

Polar Coordinates

The rectangular coordinate system is the most widely used coordinate system. Second in importance is the **polar coordinate system**. It consists of a fixed point 0 called the **pole**, or **origin**. Extending from this point is a ray called the **polar axis**. This ray usually is situated horizontally and to the right of the pole. Any point, P, in the plane can be located by specifying an angle and a distance. The angle, θ, is measured from the polar axis to a line that passes through the point and the pole. If the angle is measured in a counterclockwise direction, the angle is positive. If the angle is measured in a clockwise direction, the angle is negative. The directed

distance, *r*, is measured from the pole to point *P*. If point *P* is on the terminal side of angle θ, then the value of r is positive. If point *P* is on the opposite side of the pole, then the value of *r* is negative. The **polar coordinates** of a point can be written as an ordered pair *(r, θ)*. The location of a point can be named using many different pairs of polar coordinates. Figure 6-1 illustrates three different sets of polar coordinates for the point *P* (4,50°).

Figure 6-1 Polar forms of coterminal angles.

Conversion between polar coordinates and rectangular coordinates is illustrated as follows and in Figure 6-2.

$$r^2 = x^2 + y^2$$
$$\frac{x}{r} = \cos\theta \Rightarrow x = r\cos\theta$$
$$\frac{y}{r} = \sin\theta \Rightarrow y = r\sin\theta$$

Figure 6-2 Polar to rectangular conversion.

Example1: Convert $P(4,9)$ to polar coordinates.

$$r^2 = x^2 + y^2$$
$$r^2 = 4^2 + 9^2$$
$$r^2 = 97$$
$$r = \sqrt{97}$$

$$\cos\theta = \frac{x}{r}$$
$$\cos\theta = \frac{4}{\sqrt{97}}$$
$$\cos\theta \approx 0.406$$
$$\theta \approx 66°$$

The polar coordinates for $P(4, 9)$ are $P\left(\sqrt{97}, 66°\right)$.

Example 2: Convert $P(5,20°)$ to rectangular coordinates.

$$x = r\cos\theta$$
$$x = 5\cos 20°$$
$$x \approx (5)(0.94)$$
$$x \approx 4.7$$
$$y = r\sin\theta$$
$$y = 5\sin 20°$$
$$y \approx (5)(0.34)$$
$$y \approx 1.7$$

The rectangular coordinates for $P(5,20°)$ are $P(4.7, 1.7)$.

Example 3: Transform the equation $x^2 + y^2 + 5x = 0$ to polar coordinate form.

$$x^2 + y^2 + 5x = 0$$
$$r^2 + 5x = 0$$
$$r^2 + 5(r\cos\theta) = 0$$

$$r^2 + 5r\cos\theta = 0$$
$$r(r + 5\cos\theta) = 0$$

The equation $r = 0$ is the pole. Thus, keep only the other equation.

$$r + 5\cos\theta = 0$$

Graphs of trigonometric functions in polar coordinates are very distinctive. In Figure 6-3, several standard polar curves are illustrated. The variable a in the equations of these curves determines the size (scale) of the curve.

Figure 6-3 Graphs of some common figures in polar form.

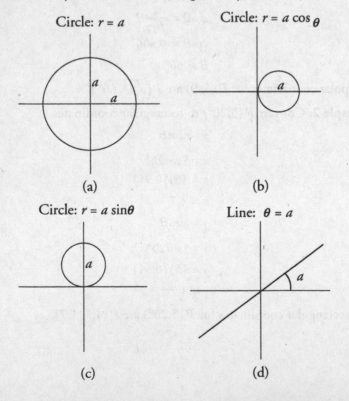

Circle: $r = a$

(a)

Circle: $r = a \cos \theta$

(b)

Circle: $r = a \sin\theta$

(c)

Line: $\theta = a$

(d)

Horizontal line: $r = a / \sin\theta$

(e)

Vertical line: $r = a / \cos\theta$

(f)

Cardioid: $r = a + a\cos\theta$

(g)

Cardioid: $r = a + a\sin\theta$

(h)

Archimedes' spiral: $r = a\theta$

(i)

Three-leaved rose: $r = a\cos 3\theta$

(j)

Three-leaved rose: $r = a \sin 3\theta$

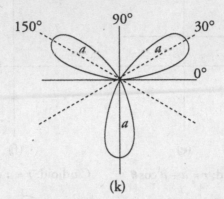

(k)

Lemniscate: $r^2 = a^2 \cos 2\theta$

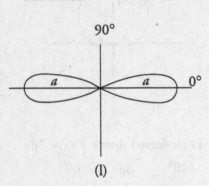

(l)

Lemniscate: $r^2 = a^2 \sin 2\theta$

(m)

Four-leaved rose: $r = a \cos 2\theta$

(n)

Four-leaved rose: $r = a \sin 2\theta$

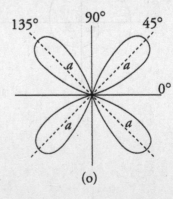

(o)

Geometry of Complex Numbers

Complex numbers can be represented in both rectangular and polar coordinates. All complex numbers can be written in the form $a + bi$, where a and b are real numbers and $i^2 = -1$. Each complex number corresponds to a point in the **complex plane** when a point with coordinates (a, b) is associated with a complex number $a + bi$. In the complex plane, the x-axis is named the **real axis** and the y-axis is named the **imaginary axis**.

Example 4: Plot $4 - 2i$, $-3 + 2i$, and $-5 - 3i$ in the complex plane (see Figure 6-4).

Figure 6-4 Complex numbers plotted in the complex plane.

Complex numbers can be converted to polar coordinates by using the relationships $x = r \cos \theta$ and $y = r \sin \theta$. Thus, if z is a complex number:

$$z = x + iy = r \cos \theta + ir \sin \theta$$
$$z = x + jy = r(\cos \theta + i \sin \theta)$$

Sometimes the expression $\cos \theta + \sin \theta$ is written as cis θ. The **absolute value**, or **modulus**, of z is $|z| = \sqrt{x^2 + y^2}$. The angle formed between the positive x-axis and a line drawn from the origin to z is called the **argument** or **amplitude** of z. If $z = x + iy$ is a complex number, then the conjugate of z is written as $\bar{z} = x - iy$.

Example 5: Convert the complex number $5 - 3i$ to polar coordinates (see Figure 6-5).

$$r = |z| = \sqrt{5^2 + (-3)^2}$$

$$r = |z| - \sqrt{34}$$

$$\cos\theta = \frac{5}{\sqrt{34}}$$

$$\cos\theta \approx 0.857$$

Reference angle $\theta \approx 31°$.

Since θ is in the fourth quadrant,

$$\theta = 360° - 31° = 329°$$

Figure 6-5 Drawing for Example 5.

Therefore,

$$5 - 3i = \sqrt{34}\left(\cos 329° + i\sin 329°\right)$$

To find the product of two complex numbers, multiply their absolute values and add their amplitudes.

$$If \quad z = a\left(\cos\alpha + i\sin\alpha\right)$$

$$and \quad w = b\left(\cos\beta + i\sin\beta\right)$$

$$then \quad zw = ab\left[\cos\left(\alpha + \beta\right) + i\sin\left(\alpha + \beta\right)\right]$$

To find the quotient of two complex numbers, divide their absolute values and subtract their amplitudes.

$$If \quad z = a(\cos\alpha + i\sin\alpha)$$

$$and \quad w = b(\cos\beta + i\sin\beta)$$

$$then \quad \frac{z}{w} = \frac{a}{b}\left[\cos(\alpha - \beta) + i\sin(\alpha - \beta)\right]$$

Example 6: If $z = a(\cos\alpha + i\sin\alpha)$ and $w = b(\cos\beta + i\sin\beta)$, then find their product zw.

$$zw = \left[a(\cos\alpha + i\sin\alpha)\right]\left[b(\cos\beta + i\sin\beta)\right]$$

$$zw = ab\left[\cos\alpha\cos\beta + i\cos\alpha\sin\beta + i\sin\alpha\cos\beta + (-1)\sin\alpha\sin\beta\right]$$

$$zw = ab\left[\cos\alpha\cos\beta - \sin\alpha\sin\beta + i(\sin\alpha\cos\beta + \cos\alpha\sin\beta)\right]$$

$$zw = ab\left[\cos(\alpha + \beta) + i\sin(\alpha + \beta)\right]$$

Example 7: If $z = a(\cos\alpha + i\sin\alpha)$ and $w = b(\cos\beta + i\sin\beta)$, then find their quotient z/w.

$$\frac{z}{w} = \frac{a(\cos\alpha + i\sin\alpha)}{b(\cos\beta + i\sin\beta)}$$

$$\frac{z}{w} = \frac{a(\cos\alpha + i\sin\alpha)(\cos\beta - i\sin\beta)}{b(\cos\beta + i\sin\beta)(\cos\beta - i\sin\beta)}$$

$$\frac{z}{w} = \frac{a}{b}\frac{\left[\cos\alpha\cos\beta - i\cos\alpha\sin\beta + i\sin\alpha\cos\beta - (-1)\sin\alpha\sin\beta\right]}{\left[\cos^2\beta - (-1)\sin^2\beta\right]}$$

$$\frac{z}{w} = \frac{a}{b}\frac{\left[(\cos\alpha\cos\beta + \sin\alpha\sin\beta) + i(\sin\alpha\cos\beta - \cos\alpha\sin\beta)\right]}{(1)}$$

$$\frac{z}{w} = \frac{a}{b}\left[\cos(\alpha - \beta) + i\sin(\alpha - \beta)\right]$$

Example 8: If $z = 4(\cos 65° + i\sin 65°)$ and $w = 7(\cos 105° + i\sin 105°)$, then find zw and z/w.

$$zw = (4)(7)\left[\cos(65° + 105°) + i\sin(65° + 105°)\right]$$

$$zw = 28(\cos 170° + i\sin 170°)$$

$$\frac{z}{w} = \frac{4}{7}\left[\cos(65° - 105°) + i\sin(65° - 105°)\right]$$

$$\frac{z}{w} = \frac{4}{7}\left[\cos\left(-40°\right) + i\sin\left(-40°\right)\right]$$

$$\frac{z}{w} = \frac{4}{7}\left[\cos\left(320°\right) + i\sin\left(320°\right)\right]$$

De Moivre's Theorem

The process of **mathematical induction** can be used to prove a very important theorem in mathematics known as **De Moivre's theorem**. If the complex number $z = r(\cos\alpha + i\sin\alpha)$, then

$$z^2\left[r\left(\cos\alpha + i\sin\alpha\right)\right]\left[r\left(\cos\alpha + i\sin\alpha\right)\right]$$

$$z^2 = r^2\left[\cos\left(\alpha + \alpha\right) + i\sin\left(\alpha + \alpha\right)\right]$$

$$z^2 = r^2\left(\cos 2\alpha + i\sin 2\alpha\right)$$

$$z^3 = r^2\left(\cos 2\alpha + i\sin 2\alpha\right)\left[r\left(\cos\alpha + i\sin\alpha\right)\right]$$

$$z^3 = r^3\left[\cos\left(2\alpha + \alpha\right) + i\sin\left(2\alpha + \alpha\right)\right]$$

$$z^3 = r^3\left(\cos 3\alpha + i\sin 3\alpha\right)$$

$$z^4 = r^3\left[\left(\cos 3\alpha + i\sin 3\alpha\right)\right]\left[r\left(\cos\alpha + i\sin\alpha\right)\right]$$

$$z^4 = r^4\left[\cos\left(\cos 3\alpha + \alpha\right) + i\sin\left(3\alpha + \alpha\right)\right]$$

$$z^4 = r^4\left(\cos 4\alpha + i\sin 4\alpha\right)$$

The preceding pattern can be extended, using mathematical induction, to De Moivre's theorem.

If $z = r(\cos\alpha + i\sin\alpha)$, and n is a natural number, then

$$z^n = r^n\left(\cos n\alpha + i\sin n\alpha\right)$$

Example 9: Write $\left(\sqrt{3} + i\right)^7$ in the form $a + bi$.

First determine the radius:

$$r = \left|\sqrt{3} + i\right|$$

$$r = \sqrt{\sqrt{3}^2 + (-1)^2}$$

$$r = \sqrt{3 + 1}$$

$$r = 2$$

Since $\cos \alpha = \sqrt{3}/2$ and $\sin \alpha = 1/2$, α must be in the first quadrant and $\alpha = 30°$. Therefore,

$$\left(\sqrt{3} + i\right)^7 = \left[2\left(\cos 30° + i\sin 30°\right)\right]^7$$

$$\left(\sqrt{3} + i\right)^7 = 2^7\left[\cos\left(7 \cdot 30°\right) + i\sin\left(7 \cdot 30°\right)\right]$$

$$\left(\sqrt{3} + i\right)^7 = 128\left(\cos 210° + i\sin 210°\right)$$

$$\left(\sqrt{3} + i\right)^7 = 128\left(-\frac{\sqrt{3}}{2} + \frac{-1}{2}i\right)$$

$$\left(\sqrt{3} + i\right)^7 = 128\left(-\frac{\sqrt{3}}{2} - \frac{1}{2}i\right)$$

$$\left(\sqrt{3} + i\right)^7 = -64\sqrt{3} - 64i$$

Example 10: Write $\left(\sqrt{2} - i\sqrt{2}\right)^4$ in the form $a + bi$.

First determine the radius:

$$r = \left|\sqrt{2} - i\sqrt{2}\right|$$

$$r = \sqrt{\sqrt{2}^2 + \left(-\sqrt{2}\right)^2}$$

$$r = \sqrt{2 + 2}$$

$$r = 2$$

Since $\cos \alpha = \sqrt{2}/2$ and $\sin \alpha = -\sqrt{2}/2$, α must be in the fourth quadrant and $\alpha = 315°$. Therefore,

$$\left(\sqrt{2} - i\sqrt{2}\right)^4 = \left[2\left(\cos 315° + i\sin 315°\right)\right]^4$$

$$\left(\sqrt{2} - i\sqrt{2}\right)^4 = 2^4\left[\cos\left(4 \cdot 315°\right) + i\sin\left(4 \cdot 315°\right)\right]$$

$$\left(\sqrt{2} - i\sqrt{2}\right)^4 = 16\left(\cos 1260° + i\sin 1260°\right)$$

$$\left(\sqrt{2} - i\sqrt{2}\right)^4 = 16\left(\cos 180° + i\sin 180°\right)$$

$$\left(\sqrt{2} - i\sqrt{2}\right)^4 = 16(-1 + 0i)$$

$$\left(\sqrt{2} - i\sqrt{2}\right)^4 = -16 + 0i$$

$$\left(\sqrt{2} - i\sqrt{2}\right)^4 = -16$$

Problems involving powers of complex numbers can be solved using binomial expansion, but applying De Moivre's theorem is usually more direct.

De Moivre's theorem can be extended to roots of complex numbers yielding the **nth root theorem.** Given a complex number $z = r(\cos \alpha + i \sin\alpha)$, all of the *n*th roots of z are given by

$$r^{1/n}\left[\cos\left(\frac{\alpha + k \cdot 360°}{n}\right) + i\sin\left(\frac{\alpha + k \cdot 360°}{n}\right)\right]$$

where $k = 0, 1, 2, \ldots, (n-1)$

If $k = 0$, this formula reduces to

$$r^{1/n}\left[\cos\left(\frac{\alpha}{n}\right) + i\sin\left(\frac{\alpha}{n}\right)\right]$$

This root is known as the **principal nth root** of z. If $\alpha = 0°$ and $r = 1$, then $z = 1$ and the **nth roots of unity** are given by

$$r^{1/n}\left[\cos\left(\frac{k \cdot 360°}{n}\right) + i\sin\left(\frac{k \cdot 360°}{n}\right)\right]$$

where $k = 0, 1, 2, \ldots, (n-1)$

Example 11: What are each of the five fifth-roots of $z = \sqrt{3} + i$ expressed in trigonometric form?

$$r = \left|\sqrt{3} + i\right|$$
$$r = \sqrt{\sqrt{3}^2 + (-1)^2}$$
$$r = \sqrt{3+1}$$
$$r = 2$$

Since $\cos\alpha = \sqrt{3}/2$ and $\sin\alpha = 1/2$, α is in the first quadrant and $\alpha = 30°$. Therefore, since the sine and cosine are periodic,

$$z = r\left(\cos\alpha + i\sin\alpha\right)$$
$$z = 2\left[\cos\left(30° + k \cdot 360°\right) + i\sin\left(30° + k \cdot 360°\right)\right]$$

and applying the *n*th root theorem, the five fifth-roots of z are given by

and applying the nth root theorem, the five fifth-roots of z are given by

$$2^{1/5}\left[\cos\left(\frac{30° + k \cdot 360}{5}\right) + i\sin\left(\frac{30° + k \cdot 360}{5}\right)\right]$$

where $k = 0, 1, 2, 3$, and 4

Thus the five fifth-roots are

$$z_1 = 2^{1/5}\left(\cos 6° + i\sin 6°\right)$$

$$z_2 = 2^{1/5}\left(\cos 78° + i\sin 78°\right)$$

$$z_3 = 2^{1/5}\left(\cos 150° + i\sin 150°\right)$$

$$z_4 = 2^{1/5}\left(\cos 222° + i\sin 222°\right)$$

$$z_5 = 2^{1/5}\left(\cos 294° + i\sin 294°\right)$$

Observe the even spacing of the five roots around the circle in Figure 6-6.

Figure 6-6 Drawing for Example 11.

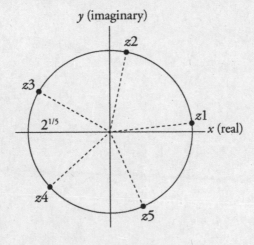

Chapter Checkout

Q&A

1. Convert P(2, 5) from rectangular coordinates to polar coordinates.

2. Convert P(8, 26°) from polar coordinates to rectangular coordinates.

3. True or False: The Rose: r = α sin 2 θ has two "petals."

4. Convert the complex number 4 + 2*i* to polar coordinates.

5. If **z** = 2(cos70° + *i* sin 70°) and **w**=6(cos80° + *i* sin 80°), then find **zw**.

Answers: 1. $r = \sqrt{29}$; $\theta = 68.2°$ **2.** P(7.2, 3.5) **3.** F
4. $\sqrt{20}(\cos 26.6° + i \sin 26.6°)$ **5.** 12(cos 150° + *i* sin 150°).

Chapter 7

INVERSE FUNCTIONS
AND EQUATIONS

Chapter Checkin

❑ Defining inverse trig functions

❑ Demonstrating how to restrict the basic trig functions to certain quadrants

❑ Showing that the restricted trig functions are one to one and have inverses

❑ Solving problems using inverse trig functions

❑ Identifying trig equations with primary solutions

❑ Solving trig equations

The standard trig functions are periodic, meaning that they repeat themselves. Therefore, the same output value appears for multiple input values of the function. This makes inverse functions impossible to construct. In order to solve equations involving trig functions, it is imperative for inverse functions to exist. Thus, mathematicians have to restrict the trig function in order create these inverses.

Inverse Cosine and Inverse Sine

To define an inverse function, the original function must be **one-to-one**. For a one-to-one correspondence to exist, (1) each value in the domain must correspond to exactly one value in the range, and (2) each value in the range must correspond to exactly one value in the domain. The first restriction is shared by all functions; the second is not. The sine function, for example, does not satisfy the second restriction, since the same value in the range corresponds to many values in the domain (see Figure 7-1).

Figure 7-1 Sine function is not one to one.

To define the inverse functions for sine and cosine, the domains of these functions are restricted. The restriction that is placed on the domain values of the cosine function is $0 \leq x \leq \pi$ (see Figure 7-2). This restricted function is called Cosine. Note the capital "C" in Cosine.

Figure 7-2 Graph of restricted cosine function.

Domain: $0 \leq x \leq \pi$

Range: $-1 \leq y \leq 1$

The **inverse cosine function** is defined as the inverse of the restricted Cosine function $\mathrm{Cos}^{-1}(\cos x) = x \quad 0 \leq x \leq \pi$. Therefore,

$$y = \mathrm{Cos}^{-1}x, \text{ where } 0 \leq y \leq \pi \text{ and } -1 \leq x \leq 1$$

Figure 7-3 Graph of inverse cosine function.

Domain: $-1 \leq x \leq 1$

Range: $0 \leq y \leq \pi$

Identities for the cosine and inverse cosine:

$$\cos(\text{Cos}^{-1}x) = x \qquad -1 \leq x \leq 1$$

$$\text{Cos}^{-1}(\cos x) = x \qquad 0 \leq x \leq \pi$$

The inverse sine function's development is similar to that of the cosine. The restriction that is placed on the domain values of the sine function is

$$-\frac{\pi}{2} \leq x \leq \frac{\pi}{2}$$

This restricted function is called Sine (see Figure 7-4). Note the capital "S" in Sine.

Figure 7-4 Graph of restricted sine function.

Domain: $-\pi/2 \leq x \leq \pi/2$

Range: $-1 \leq y \leq 1$

The **inverse sine function** (see Figure 7-5) is defined as the inverse of the restricted Sine function $y = \text{Sin } x$,

$$-\frac{\pi}{2} \le x \le \frac{\pi}{2}$$

Therefore,

$$-\frac{\pi}{2} \le x \le \frac{\pi}{2} \text{ and } y = \text{Sin}^{-1}x, \text{ where} -\frac{\pi}{2} \le y \le \frac{\pi}{2} \text{ and} -1 \le x \le 1$$

Figure 7-5 Graph of inverse sine function.

Domain: $-1 \le x \le 1$

Range: $-\pi/2 \le y \le \pi/2$

Identities for the sine and inverse sine:

$$\text{Sin}\left(\sin^{-1}x\right) = x \ -1 \le x \le 1$$

$$\text{Sin}^{-1}(\sin x) = x \ -\frac{\pi}{2} \le x \le \frac{\pi}{2}$$

The graphs of the functions $y = \text{Cos } x$ and $y = \text{Cos}^{-1} x$ are reflections of each other about the line $y = x$. The graphs of the functions $y = \text{Sin} x$ and $y = \text{Sin}^{-1}x$ are also reflections of each other about the line $y = x$ (see Figure 7-6).

Figure 7-6 Symmetry of inverse sine and cosine.

(a) (b)

Example 1: Using Figure 7-7, find the exact value of $\text{Cos}^{-1}(-\sqrt{3}/2)$.

If $y = \text{Cos}^{-1}\left(\dfrac{-\sqrt{3}}{2}\right)$, then $\cos y = \dfrac{-\sqrt{3}}{2}$ where $0 \le y \le \pi$

Thus, $y = 5\pi/6$ or $y = 150°$.

Figure 7-7 Drawing for Example 1.

Example 2: Using Figure 7-8, find the exact value of $\text{Sin}^{-1}\left(\sqrt{2}/2\right)$

If $y = \text{Sin}^{-1}\left(\dfrac{\sqrt{2}}{2}\right)$, then $\sin y = \dfrac{\sqrt{2}}{2}$ where $-\dfrac{\pi}{2} \le x \le \dfrac{\pi}{2}$

Figure 7-8 Drawing for Example 2.

Thus, $y = \pi/4$ or $y = 45°$.

Example 3: Find the exact value of $\cos\left(\text{Cos}^{-1}\,0.62\right)$.

Use the cosine-inverse cosine identity:

$$\cos\left(\text{Cos}^{-1}\,0.62\right) = 0.62$$

Other Inverse Trigonometric Functions

To define the inverse tangent, the domain of the tangent must be restricted to

$$-\frac{\pi}{2} < x < \frac{\pi}{2}$$

This restricted function is called Tangent (see Figure 7-9). Note the capital "T" in Tangent.

Figure 7-9 Graph of restricted tangent function.

The **inverse tangent function** (see Figure 7-10) is defined as the inverse of the restricted Tangent function $y = \text{Tan } x$,

$$-\frac{\pi}{2} < x < \frac{\pi}{2}$$

Therefore,

$$y = \text{Tan}^{-1} x, \text{ where } -\frac{\pi}{2} < y < \frac{\pi}{2} \text{ and } -\infty < x < \infty$$

Figure 7-10 Graph of inverse tangent function.

Domain: All real numbers

Range: $-\frac{\pi}{2} < y < \frac{\pi}{2}$

Identities for the tangent and inverse tangent:

$$\tan\left(\text{Tan}^{-1} x\right) = x \quad -\infty < x < \infty$$

$$\text{Tan}^{-1}(\tan x) = x \quad -\frac{\pi}{2} < x < \frac{\pi}{2}$$

The **inverse tangent, inverse secant**, and **inverse cosecant** functions are derived from the restricted Sine, Cosine, and Tangent functions. The graphs of these functions are shown in Figure 7-11.

Figure 7-11 Graphs of inverse cotangent, inverse secant, and inverse cosecant functions.

Domain: All real numbers
Range: $0 < y < \pi$

(a)

Domain: $x \leq -1$ or $x \geq 1$
Range: $0 \leq y \leq \pi, y \neq \pi/2$

(b)

Domain: $x \leq -1$ or $x \geq 1$
Range: $-\pi/2 \leq y \leq \pi/2, y \neq 0$

(c)

Trigonometric identities involving inverse cotangent, inverse secant, and inverse cosecant:

$$\text{Cot}^{-1} x = \text{Tan}^{-1} \frac{1}{x} \qquad \text{where } x > 0$$

$$\text{Cot}^{-1} x = \pi + \text{Tan}^{-1} \frac{1}{x} \qquad \text{where } x < 0$$

$$\text{Sec}^{-1} x = \text{Cos}^{-1} \frac{1}{x} \qquad \text{where } x \geq 1 \text{ or } x \leq -1$$

$$\text{Csc}^{-1} x = \text{Sin}^{-1} \frac{1}{x} \qquad \text{where } x \geq 1 \text{ or } x \leq -1$$

Example 4: Determine the exact value of $\sin\left[\operatorname{Sec}^{-1}(-4)\right]$ without using a calculator or tables of trigonometric functions.

$$\text{If } \alpha = \operatorname{Sec}^{-1}(-4), \text{ then}$$

$$\operatorname{Sec}\alpha = -4, \text{ where } 0 \le \alpha \le \pi, \ \alpha \ne \pi/2$$

In this range, the cosine and the secant are negative in the second quadrant. From this reference triangle, calculate the third side and find the sine (see Figure 7-12).

$$a^2 + (-1)^2 = 4^2$$
$$a^2 = 4^2 - (-1)^2$$
$$a = \sqrt{16 - 1}$$
$$a = \sqrt{15}$$

Figure 7-12 Drawing for Example 4.

Therefore,

$$\sin\left(\operatorname{Sec}^{-1}(-4)\right) = \sin\alpha$$
$$= \frac{a}{4}$$
$$= \frac{\sqrt{15}}{4}$$

Example 5: Determine the exact value of cos (Tan^{-1}7) without using a calculator or tables of trigonometric functions.

<p style="text-align:center">If α = Tan^{-1}7, then</p>

$$\text{Tan } \alpha = 7, \text{ where } -\frac{\pi}{2} < \alpha < \frac{\pi}{2}$$

In this range, the tangent and the cotangent are positive in the first quadrant. From this reference triangle, calculate the third side and find the cosine (see Figure 7-13).

$$c^2 = 1^2 + 7^2$$
$$c^2 = 50$$
$$c = \sqrt{50}$$
$$c = 5\sqrt{2}$$

Figure 7-13 Drawing for Example 5.

Therefore,

$$\cos\left(\text{Tan}^{-1}7\right) = \cos\alpha$$
$$= \frac{1}{5\sqrt{2}}$$
$$= \frac{\sqrt{2}}{10}$$

Trigonometric Equations

Trigonometric identities are true for all replacement values for the variables for which both sides of the equation are defined. **Conditional trigonometric equations** are true for only some replacement values. Solutions in a specific interval, such as $0 \leq x \leq 2\pi$, are usually called **primary solutions**. A **general solution** is a formula that names all possible solutions.

The process of solving general trigonometric equations is not a clear-cut one. No rules exist that will always lead to a solution. The procedure usually involves the use of identities, algebraic manipulation, and trial and error. The following guidelines can help lead to a solution.

If the equation contains more than one trigonometric function, use identities and algebraic manipulation (such as factoring) to rewrite the equation in terms of only one trigonometric function. Look for expressions that are in quadratic form and solve by factoring. Not all equations have solutions, but those that do usually can be solved using appropriate identities and algebraic manipulation. Look for patterns. There is no substitute for experience.

Example 6: Find the exact solution:

$$\cos^2 \alpha = -\cos\alpha + \sin^2 \alpha, \ 0° \leq \alpha \leq 360°$$

First, transform the equation by using the identity $\sin^2 \alpha + \cos^2 \alpha = 1$.

$$\cos^2 \alpha = -\cos\alpha + \left(1 - \cos^2 \alpha\right)$$
$$2\cos^2 \alpha + \cos\alpha - 1 = 0$$
$$(2\cos\alpha - 1)(\cos\alpha + 1) = 0$$

Therefore,

$$2\cos\alpha - 1 = 0 \qquad\qquad \cos\alpha + 1 = 0$$
$$2\cos\alpha = 1 \qquad\qquad \cos\alpha = -1$$
$$\cos\alpha = 1/2 \qquad\qquad \alpha = 180°$$
$$\alpha = 60°, 300°$$

Thus,

$$\alpha = 60°, 180°, 300°$$

Example 7: Solve cos 2x = 3(sin x − 1) for all real values of x.

$$\cos 2x = 3\left(\sin x - 1\right) \qquad \text{given}$$

$$1 - 2\sin^2 x = 3\sin x - 3 \qquad \text{double angle formula}$$

$$2\sin^2 x + 3\sin x - 4 = 0 \qquad \text{quadratic equation}$$

$$\sin x = \frac{-3 \pm \sqrt{9 - (4)(2)(-4)}}{(2)(2)} \qquad \text{use quadratic formula}$$

$$\sin x = \frac{-3 \pm \sqrt{41}}{4}$$

$$\sin x = -2.351 \ or \ 0.8508$$

The first answer, −2.351, is not a solution, since the sine function must range between − 1 and 1. The second answer, 0.8508, is a valid value. Thus, if *k* is an integer,

$$x = \text{Sin}^{-1} 0.8508 + 2k\pi \quad x = \pi - \text{Sin}^{-1} 0.8508 + 2k\pi$$

In radian form,

$$x = 1.0175 + 2k\pi \quad x = 2.124 + 2k\pi$$

In degree form,

$$x = 58.3° + \left(360°\right)(k) \quad x = 121.7° + \left(360°\right)(k)$$

Example 8: Find the exact solution:

$$\cos 2\theta = \cos\theta, 0° < \theta < 180°$$

First, transform the equation by using the double angle identity cos 2θ = 2 cos²θ − 1.

$$\cos 2\theta = \cos\theta$$

$$2\cos^2\theta - 1 = \cos\theta$$

$$2\cos^2\theta - \cos\theta - 1 = 0$$

$$\left(2\cos\theta + 1\right)\left(\cos\theta - 1\right) = 0$$

Therefore,

$$2\cos\theta + 1 = 0 \qquad \cos\theta - 1 = 0$$
$$2\cos\theta = -1 \qquad \cos\theta = 1$$
$$\cos\theta = -\tfrac{1}{2} \qquad \theta = 0°$$
$$\theta = 120°$$

Thus,

$$\theta = 0°, 120°$$

Chapter Checkout

Q&A

1. Solve $\sin\theta = 2\cos^2\theta - 1$ where $0° \leq \theta < 360°$.
2. Find the exact value of $\cos(\sin^{-1}(-\tfrac{3}{7}))$.
3. Find the exact value of $\tan(\sin^{-1}\tfrac{2}{3})$.
4. True or False: The inverse sine and inverse cosine are defined in the same quadrants.
5. Find the exact value of $\sin(\sin^{-1}\tfrac{3}{8})$.
6. Solve $\sin^2 2\theta = 1$.

Answers: 1. 30°, 150°, 270° **2.** $\dfrac{2\sqrt{10}}{7}$ **3.** $\dfrac{2\sqrt{5}}{5}$ **4.** F **5.** $\dfrac{3}{8}$ **6.** 45°, 225°.

Chapter 8
ADDITIONAL TOPICS

Chapter Checkin

❑ Defining an alternate form for the equation y = A sin (Bt + C)

❑ Converting between y = A sin (Bt + C) and y = M sin Bt + N cos Bt

❑ Finding period, frequency, and phase shift

❑ Defining uniform circular motion

❑ Using uniform circular motion to solve problems about linear velocity

❑ Defining simple harmonic motion in terms of uniform circular motion

❑ Solving problems using simple harmonic motion

Have you ever noticed that the motions of some objects seem to be very rhythmic and repetitive? Motions like train wheels and linkages, a child's swing, the pistons in a car engine, throwing a ball, and bouncing a ball are all related to harmonic motion and the sine curve. Understanding simple harmonic motion and uniform circular motion can help explain how most of these very common movements are related.

The Expression *M* sin *Bt* + *N* cos *Bt*

The equation y = *M* sin *Bt* + *N* cos *Bt* and the equation y = *A* sin *(Bt + C)* are equivalent where the relationships of *A, B, C, M,* and *N* are as follows. The proof is direct and follows from the sum identity for sine. The following is a summary of the properties of this relationship.

$M \sin Bt + N\cos Bt = \sqrt{M^2 + N^2} \sin (Bt + C)$ given that *C* is an angle with a point *P(M, N)* on its terminal side (see Figure 8-1).

$$\sin C = \frac{N}{\sqrt{M^2 + N^2}}$$

$$\cos C = \frac{M}{\sqrt{M^2 + N^2}}$$

$$\text{amplitude} = \sqrt{M^2 + N^2}$$

$$\text{period} = \frac{2\pi}{B}$$

$$\text{frequency} = \frac{B}{2\pi}$$

$$\text{phase shift} = -\frac{C}{B}$$

Figure 8-1 Reference graph for $y = M \sin Bt + N \cos Bt$.

Example 1: Convert the equation $y = \sqrt{5} \sin 3t + 2 \cos 3t$ to the form $y = A \sin (Bt + C)$. Find the period, frequency, amplitude, and phase shift (see Figure 8-2).

$$M = \sqrt{5}$$

$$N = 2$$

$$B = 3$$

$$R = \sqrt{M^2 + N^2} = 3$$

$$\sin C = \frac{2}{3}$$

$$\cos C = \frac{\sqrt{5}}{3}$$

$$C \approx 0.7297$$

$$y = 3 \sin (3t + 0.7297)$$

$$\text{amplitude} = \sqrt{M^2 + N^2} = 3$$

$$\text{period} = \frac{2\pi}{B} = \frac{2\pi}{3} \approx 2.094$$

$$\text{frequency} = \frac{B}{2\pi} = \frac{3}{2\pi} \approx 0.477$$

$$\text{phase shift} = -\frac{C}{B} \approx -\frac{0.7297}{3} \approx -0.2432$$

Figure 8-2 Drawing for Example 1.

Example 2: Convert the equation $y = -\sin \pi t + \cos \pi t$ to the form $y = A \sin (Bt + C)$. Find the period, frequency, amplitude, and phase shift (see Figure 8-3).

$$M = -1$$

$$N = 1$$

$$B = \pi$$

$$R = \sqrt{M^2 + N^2} = \sqrt{2}$$

$$\sin C = \frac{\sqrt{2}}{2}$$

$$\cos C = \frac{-\sqrt{2}}{2}$$

$$C = \frac{3\pi}{4}$$

$$y = \sqrt{2} \sin\left(\pi t + \frac{3\pi}{4}\right)$$

$$\text{amplitude} = \sqrt{M^2 + N^2} = \sqrt{2}$$

$$\text{period} = \frac{2\pi}{B} = \frac{2\pi}{\pi} \approx 2$$

$$\text{frequency} = \frac{\pi}{2\pi} = \frac{1}{2}$$

$$\text{phase shift} = -\frac{C}{B} = -\frac{\frac{3\pi}{4}}{\pi} \approx -\frac{3}{4}$$

Figure 8-3 Drawing for Example 2.

Uniform Circular Motion

If α is the measure of a central angle of a circle, measured in radians, then the length of the intercepted arc (s) can be found by multiplying the radius of the circle (r) by the size of the central angle (α); $s = r\,\alpha$. Remember, α must be measured in radians.

Example 3: Find the length (s) of the arc intercepted by a central angle of size 3 radians if the radius of the circle is 5 centimeters (see Figure 8-4).

$$s = r\,\alpha$$
$$s = (5)(3)$$
$$s = 15$$

Figure 8-4 Drawing for Example 3.

Thus, the length of the intercepted arc is 15 centimeters.

Example 4: Using Figure 8-5, find the length (s) of the arc intercepted by a central angle of size $- 100°$ if the radius of the circle is 7 centimeters.

Round the answer to two decimal places. (In this problem, the negative value of the angle and the arc length refer to a negative direction.)

First, convert $-100°$ to radian measure.

$$\alpha - \left(\frac{\pi}{180}\right)(-100) \qquad s = \alpha r$$
$$\alpha = 1.75 \qquad\qquad s = (1.75)(7)$$
$$s = -12.25$$

Figure 8-5 Drawing for Example 4.

Thus, the length of the intercepted arc is −12.25 centimeters.

The linear velocity *(v)* of a point traveling at a constant speed along an arc of a circle is given as:

$$v = \frac{\text{length of the arc}}{\text{time}} = \frac{r\,\alpha}{t}$$

Example 5: If the earth has a radius of 4,050 miles and rotates one complete revolution *(2π radians)* each 24 hours, what is the linear velocity of an object located on the equator?

$$v = \frac{r\,\alpha}{t}$$
$$v = \frac{(4050)(2\pi)}{24}$$
$$v \approx \frac{(4050)(6.28)}{24}$$
$$v \approx 1060$$

Thus, the linear velocity of the object is 1,060 miles per hour.

The angular velocity (ω) of a point traveling at a constant speed along an arc of a circle is given as:

$$\omega = \frac{\text{measure of angle of rotation}}{\text{time}} = \frac{\alpha}{t}$$

Angular and linear velocity are both positive if the movement is counterclockwise and negative if the movement is clockwise.

Example 6: Point P revolves counterclockwise around a point O making 7 complete revolutions in 5 seconds. If the radius of the circle shown in Figure 8-6 is 8 centimeters, find the linear and angular velocities of point P. Approximate π to two decimal places.

$$v = \frac{r\alpha}{t}$$

$$v \approx \frac{(8)(7)(2\pi)}{5}$$

$$v = \frac{(8)(7)(6.28)}{5}$$

$$v \approx 70.34$$

Figure 8-6 Drawing for Example 6.

Thus, the linear velocity is approximately 70.34 centimeters per second.

$$\omega = \frac{\alpha}{t}$$

$$\omega = \frac{(7)(2\pi)}{5}$$

$$\omega \approx \frac{(7)(6.28)}{5}$$

$$\omega \approx 8.79$$

Thus, the angular velocity is approximately 8.79 radians per second

Simple Harmonic Motion

Circular functions representing periodic motion that satisfy the equations

$$d = A\sin Bt \text{ and } d = A\cos Bt$$

where d is an amount of displacement, A and B are constants determined by the specific motion, and t is a measurement of time are referred to as **simple harmonic motion**.

Example 7: If the instantaneous voltage in a current is given by the equation $E = 204 \sin 3680t$, where E is expressed in volts and t is expressed in seconds, find E if $t = 0.27$ seconds. Use 3.1416 for π.

$$E = 204\sin 3680t$$

$$E = 204\sin\left[(3680)(0.27)\right]$$

$$E = 204\sin 993.6$$

Since $\sin x = \sin(x - 2k\pi)$ and $(993.6 \div 2\pi) = 158$ with a remainder of .8544,

$$E = 204\sin\left[993.6 - (158)(2\pi)\right]$$

$$E = 204\sin\left[993.6 - (158)(6.2832)\right]$$

$$E = 204\sin 0.8544$$

$$E \approx (204)(0.7542)$$

$$E \approx 153.86$$

Example 8: The horizontal displacement (d) of the end of a pendulum is $d = K\sin 2\pi t$. Find K if $d = 12$ centimeters and $t = 3.25$ seconds.

$$d = K\sin 2\pi t$$

$$12 \approx K\sin\left[(2)(3.1416)(3.25)\right]$$

$$12 \approx K\sin 20.42$$

$$K \approx \frac{12}{\sin 20.41}$$

$$K \approx \frac{12}{1}$$

$$K \approx 12$$

Chapter Checkout

Q&A

1. Given the equation y = 3sin 4t + 6 cos4t, find the amplitude and the period of the function.

2. Find the length of an arc intercepted by a central angle of size 2.3 radians if the circle has a radius of 12 inches.

3. If a ball of radius 2 feet is spinning at 12 rpm, what is the linear velocity of a point on the equator of the ball?

4. If a point revolves around a circle of radius 12 at a constant rate of 4 revolutions every 2 minutes, find its angular velocity.

5. If the displacement of a spring (d) is given by d = A cosαt where A, the initial displacement and d are expressed in inches, t in seconds, and α = 8, find d when A is 10 inches and t = 2π.

Answers: 1. Amplitude = $3\sqrt{5}$ period = $\frac{\pi}{2}$ **2.** 27.6 inches **3.** 48 π feet/min
4. $\omega = \frac{\alpha}{t} = \frac{4(2\pi)}{2} = 4\pi$ **5.** 10 inches.

CQR REVIEW

Use this CQR Review to practice what you've learned in this book. After you work through the review questions, you're well on your way to achieving your goal of understanding trigonometry.

Chapter 1

1. Convert 44.4714 to DM'S" form.

2. Find the exact value of cos270°. Do not use a calculator.

3. Find the exact value of cos60° cos 60 °. Do not use a calculator.

4. Determine the sign of the following trigonometric functions: a) sin 255° b) tan 240° c) cos (−110°).

5. Find sin θ and cos θ for the acute angle θ if $\tan \theta = \dfrac{1}{2\sqrt{2}}$.

6. What is the reference angle for −649°?

 a. 109°
 b. 19°
 c. 161°
 d. 71°

Chapter 2

7. Solve this triangle: α = 51°, β = 49°, b = 70.

8. Solve this triangle: a = 10, b = 11, α =27°.

9. Solve this triangle: b = 7, c = 4, α = 94°.

10. Solve this triangle: a = 11, b = 17, c = 14.

 a. α = 84.78°, β = 40.12°, γ = 55.10°
 b. α = 55.10°, β = 84.78°, γ = 40.12°
 c. α = 40.12°, β = 84.78°, γ = 55.10°
 d. α = 40.12°, β = 55.10°, γ = 84.78°

11. Find the area of a triangle with $a = 68°$, $b = 7$ feet, and $c = 7$ feet.

 a. 9.18 square feet
 b. 45.43 square feet
 c. 22.72 square feet
 d. 24.50 square feet

12. Find the area of a triangle with sides 4 meters, 5 meters, and 6 meters.

Chapter 3

13. For a circle of radius 3 feet, find the arc length s subtended by a central angle of $6°$.

14. Convert $171°$ to radian measure. Give the exact answer.

15. Find the exact value of $\sin \frac{\pi}{4} - \tan \frac{\pi}{4}$. Do not use a calculator.

 a. $\dfrac{2\sqrt{3} - 3}{6}$

 b. $\dfrac{1}{2}$

 c. $\dfrac{-\sqrt{3}}{2}$

 d. $\dfrac{\sqrt{2} - 2}{2}$

16. If $f(x) = \tan x$ and $f(a) = 9$, find the exact value of $f(a) + f(a + 3\pi) + f(a - 3\pi)$.

 a. 9

 b. $\dfrac{1}{9}$

 c. $\dfrac{1}{3}$

 d. 27

17. Find the exact value of $\tan\left(-\frac{2}{3}\pi\right)$.

18. Find the amplitude and period of $f(x) = -8\sin(2x)$.

Chapter 4

19. Is this statement an identity?

$$\frac{\cos x}{1 - \sin x} = \sec x + \tan x$$

20. If $\sin A = \frac{3}{7}, \frac{\pi}{2} \le A \le \pi$ and $\cos B = -\frac{2}{8}, \pi \le B \le \frac{3\pi}{2}$, find the exact value of $\cos(A + B)$.

21. Establish this identity: $\cos\left(\theta + \frac{\pi}{2}\right) = -\sin\theta$

22. Find the exact value of $\sin 2\theta$ if $\sin\theta = \frac{3}{5}, \frac{\pi}{2} < \theta < \pi$.

23. The expression $\csc 2\theta + \cot 2\theta$ forms an identity with which of the following?

 a. $\tan\theta$

 b. $\tan\frac{\theta}{2}$

 c. $\cot\theta$

 d. $\cot\frac{\theta}{2}$

24. Express $\sin 2\,\theta \cos 7\,\theta$ as a sum containing only sines or cosines.

 a. $\frac{1}{2}(\sin 9\theta + \sin 5\theta)$

 b. $\frac{1}{2}(\sin 9\theta - \sin 5\theta)$

 c. $\sin\frac{9\theta}{2} - \sin\frac{5\theta}{2}$

 d. $\sin 5\,\theta + \sin 2\theta$

Chapter 5

25. Find the position vector of the vector $v = \overrightarrow{JK}$ if $J = (5, -4)$ and $K = (-6, -7)$.

 a. $11i + 3j$
 b. $-11i + 3j$
 c. $-11i - 3j$
 d. $-2i - 11j$

26. If $v = -3i + 7j$ and $w = 5i + 10j$, find $v - w$.

 a. $-8i + 3j$
 b. $2i - 17j$
 c. $-8i + 3j$
 d. $2i + 17j$

27. If $v = 3i + 2j$ and $w = 3i + 8j$, find $5v - 3w$ and $\|5v - 3w\|$.

28. Find the unit vector having the same direction as $v = -10i + 24j$.

 a. $-\frac{5}{13}i + \frac{12}{13}j$

 b. $-\frac{10}{\sqrt{34}}i + \frac{24}{\sqrt{34}}j$

 c. $-\frac{5}{17}i + \frac{12}{17}j$

 d. $-i + j$

29. If $v = 5i - 9j$ and $w = 27i + 15j$, find $v \cdot w$.

 a. 270
 b. -450
 c. 360
 d. 0

30. Find the measure of the angle between the vectors $v = i + 4j$ and $w = -3i + 4j$.

 a. $73°$
 b. $129°$
 c. $51°$
 d. $49°$

31. Find the component form of v given $\|v\| = 6$ and the angle between the direction of v and the positive x-axis is $\alpha = 150°$.

32. Which of the following vectors is orthogonal to $-10i + 6j$?

 a. $7i + 4j$
 b. $2i + 2j$
 c. $-12i - 20j$
 d. $-15i + 9j$

Chapter 6

33. Find the rectangular coordinates of $(9, 30°)$.

a. $\left(\dfrac{9\sqrt{3}}{2}, \dfrac{9}{2} \right)$

b. $\left(\dfrac{9\sqrt{2}}{2}, \dfrac{9}{2} \right)$

c. $\left(\dfrac{9}{2}, \dfrac{9\sqrt{3}}{2} \right)$

d. $\left(-\dfrac{9\sqrt{3}}{2}, \dfrac{9}{2} \right)$

34. Find the polar coordinates of $(-12, -12)$ for $r > 0$, $0 \le \theta < 2\pi$.

35. Determine the polar form of the complex number $2 - 4i$. Express the angle θ in degrees where $0 \le \theta < 360°$, and round numerical entries to two decimal places.

Chapter 7

36. Find the exact value of $\tan\left(\cos^{-1}\left(\dfrac{1}{2} \right) \right)$.

a. $\sqrt{3}$

b. $\dfrac{\sqrt{3}}{2}$

c. $\dfrac{1}{2}$

d. $\dfrac{\sqrt{3}}{3}$

37. Solve the equation $2\sqrt{3}\cos x - 3 = 0$, where $0 \le x < 2\pi$.

38. Solve the equation $\sin^2 \theta + 2\sin \theta + 1 = 0$, where $0 \le \theta < 2\pi$.

39. Solve the equation $-\sin \theta + 1 = 2\cos^2\theta$, where $0 \le \theta < 2\pi$.

a. $\theta = 0, \theta = \pi$

b. $\theta = \dfrac{\pi}{2}, \theta = \dfrac{3\pi}{2}$

c. $\theta = 0, \theta = \dfrac{2\pi}{3}, \theta = \dfrac{4\pi}{3}$

d. $\theta = \dfrac{\pi}{2}, \theta = \dfrac{7\pi}{6}, \theta = \dfrac{11\pi}{6}$

Chapter 8

40. A propeller, 2 meters from tip to tip, is rotating at a rate of 400 revolutions per minute. Find the linear velocity of a tip of the propeller in meters per second.

41. The horizontal displacement of a pendulum is described by the equation $d = K \sin 2\pi$. Find K if $d = 20$ and $t = \frac{1}{6}$.

42. The horizontal displacement of a pendulum is described by the equation $d = K \sin 2\pi$. Find d if $t = 1.2$ and $K = 11$.

Answers: 1. $44°28'17"$ **2.** 0 **3.** $\frac{1}{4}$ **4.** negative, positive, negative **5.** $\cos\theta = \frac{2\sqrt{2}}{3}$ $\sin\theta = \frac{1}{3}$ **6.** d **7.** $\gamma = 80°$, $a = 72.08$, $c = 91.34$ **8.** $c_1 = 18.46$, $\beta_1 = 29.96°$, $\gamma_1 = 123.04°$ and $c_2 = 1.14$, $\beta_2 = 150.04°$ **9.** $a = 8.3$, $\beta = 57.3°$, $\gamma_1 = 28.7°$ **10.** c **11.** c **12.** 9.92 square meters **13.** 0.31 feet **14.** $\frac{19}{20}\pi$ **15.** d **16.** d **17.** $\sqrt{3}$ **18.** amplitude = 8 and period = π **19.** Yes **20.** $\frac{1}{56}(10\sqrt{10} + 3\sqrt{39})$ **21.** $\cos\left(\theta + \frac{\pi}{2}\right) = \cos\theta\cos\frac{\pi}{2} - \sin\theta\sin\frac{\pi}{2} =$ $(\cos\theta)(0) - \sin\theta(1) = -\sin\theta$ **22.** $-\frac{24}{25}$ **23.** c **24.** b **25.** c **26.** a **27.** $5v - 3w = 6i - 14j$, $\|5v - 3w\| = 2\sqrt{58}$ **28.** a **29.** d **30.** c **31.** $-3\sqrt{3}i + 3j$ **32.** c **33.** a **34.** $\left(12\sqrt{2}, \frac{5\pi}{4}\right)$ **35.** $4.47(\cos 296.57° + i\sin 296.57°)$ **36.** a **37.** $\left(\frac{\pi}{6}, \frac{11\pi}{6}\right)$ **38.** $\theta = \frac{3\pi}{2}$ **39.** d **40.** 800π meters per minute **41.** $\frac{40\sqrt{3}}{3}$ **42.** 10.45.

CQR RESOURCE CENTER

CQR Resource Center offers the best resources available in print and online to help you study and review the core concepts of trigonometry. You can find additional resources, plus study tips and tools to help test your knowledge, at www.cliffsnotes.com.

Books

This CliffsQuickReview book is one of many great books about trigonometry. If you want some additional resources, check out these other publications:

Schaum's Outline of Trigonometry, by Robert E. Moyer and Frank Ayres, focuses on plane trigonometry and contains hundreds of problems. McGraw-Hill Professional Book Group.

Trigonometry the Easy Way, by Douglas A. Downing, takes the form of a fantasy novel where the King of Carmorra and his subjects solve practical problems by applying principles of trigonometry. Barron's Educational Series, Inc.

Houghton Mifflin Harcourt also has two Web sites that you can visit to read about all the books we publish:

■ www.cliffsnotes.com

■ www.dummies.com

Internet

The Internet is loaded with web sites related to trigonometry. Many offer tutorials. Some link to many other useful sites about trigonometry and other related subjects. Just spend some time exploring, and you will probably find what you need. In particular, visit the following Web sites for more information about trigonometry:

Dave's short course in trigonometry, aleph0.clarku.edu/~djoyce/ java/trig/, introduces you to trigonometry and has a few exercises.

An introduction to Trigonometry, www.ping.be/math/, provides a more advanced tutorial on trigonometry.

Syvum Homepage: Online Education and Interactive Learning, www.syvum.com/math/trigonometry.html, offers a variety of math problems, including some on trigonometry.

SOSMath Homepage, www.sosmath.com/trig/trig.html, is a valuable resource that lists hundreds of sites covering all phases of mathematics.

Next time you're on the Internet, don't forget to drop by www.cliffsnotes.com.

GLOSSARY

AAS reference to solving a triangle given the measure of two angles and the length of a non-included side.

absolute value of a complex number square root of the sum of the squares of its real and imaginary coefficients.

algebraic vector an ordered pair of numbers representing the terminal point of a standard vector.

amplitude of a complex number same as the argument of a complex number.

amplitude the vertical stretch of a function.

angle a measure of rotation.

angle of depression an angle measured below the horizontal.

angle of elevation an angle measured above the horizontal.

angular velocity defined in terms of angle of rotation and time.

argument of a complex number angle formed between the positive x-axis and a line segment between the origin and the number.

ASA reference to solving a triangle given the measure of two angles and the length of the included side.

ASTC an acronym representing which trigonometric functions are positive in the I, II, III, and IV quadrants respectively.

asymptotes lines representing undefined values for trigonometric functions.

bearing an angle measured clockwise from due north to a vector.

circular functions functions whose domains are angles measured in radians and whose ranges are values that correspond to analogous trigonometric functions.

cofunction identities fundamental identities that involve the basic trig functions of complementary angles.

cofunctions pairs of trigonometric functions of complimentary angles whose trigonometric ratios are equal.

complex plane a coordinate system for complex numbers.

component vectors the horizontal and vertical component vectors of a given vector.

components the individual vectors that are combined to yield the resultant vector.

components of an algebraic vector the ordered pair of numbers representing the vector.

conditional equation an equation that is valid for a limited number of values of the variable.

conditional trigonometric equations true for only a limited number of replacement values.

conjugate of a complex number same as original except for the sign of the imaginary component.

cosecant the reciprocal of the sine function.

cosine a trigonometric ratio equal to the adjacent side divided by the hypotenuse.

cotangent the reciprocal of the tangent function.

coterminal two angles in standard position that share a terminal side.

De Moivre's theorem a theorem involving powers of complex numbers.

degree a unit of angle measurement equal to 1/360 of a revolution.

difference identities for tangent identities involving the tangents of differences of angles.

difference identity for cosine one of the trigonometric addition identities.

difference identity for sine one of the trigonometric addition identities.

directed line segment a line segment of a given length and a given direction.

dot product a process of combining two vectors yielding a single number.

double-angle identities useful in writing trig functions involving double angles as trig functions of single angles.

double-angle identities for tangent useful in writing trig functions involving double angles as functions of single numbers.

equivalent vectors two vectors that have the same magnitude and direction.

even function a function is even if $f(-x) = f(x)$.

general solution solutions defined over entire domain.

geometric vector a quantity that can be represented by a directional line segment.

half-angle identities useful in writing trig functions involving half angles as trig functions of single angles.

half-angle identities for tangent useful in writing trig functions involving half angles as functions of single angles.

Heron's formula a formula for finding the area of a triangle given the lengths of the three sides.

identities for negatives fundamental identities that involve the basic trig functions of negative angles.

identity see trigonometric identity.

imaginary axis an axis in the complex plane.

initial point the beginning point of a vector.

initial side side of angle where angle measurement begins.

inverse cosecant function defined in terms of the restricted sine function.

inverse cosine function inverse of the restricted cosine function.

inverse cotangent function defined in terms of the restricted tangent function.

inverse notation notation used to express an angle in terms of the value of trigonometric functions.

inverse secant function defined in terms of the restricted cosine function.

inverse sine function inverse of the restricted sine function.

inverse tangent function inverse of the restricted tangent function.

law of cosines a relationship between the lengths of the three sides of a triangle and the cosine of one of the angles.

law of sines a relationship between the ratios of the sines of angles of a triangle and the side opposite those angles.

linear interpolation a method of approximating values in a table using adjacent table values.

linear velocity defined in terms of arc length and time.

magnitude of a vector the length of the directional line segment.

mathematical induction a method of mathematical proof.

maximum value largest value of a function in a given interval.

minimum value smallest value of a function in a given interval.

minute an angle measurement equal to 1/60 of a degree.

modulus of a complex number same as absolute value of a complex number.

negative angle results from clockwise rotation.

norm another name for the magnitude of a vector.

nth root theorem an extension of De Moivre's theorem involving roots of complex numbers.

odd function a function is odd if $f(-x) = -f(x)$.

odd-even identities see identities for negatives.

one-to-one a characteristic of functions where each element in the domain is pairs with one and only one element in the range and vice versa.

orthogonal perpendicular.

parallelogram rule a process used to add together two nonparallel vectors.

period the smallest value of q such that $f(x) = f(x+q)$ where $f(x)$ is a periodic function.

periodic functions trigonometric functions whose values repeat once each period.

phase shift the horizontal displacement of a function to the right or left of the vertical axis.

polar axis a ray extending from the pole in a polar coordinate system.

polar coordinate system a coordinate system using distance and angle for position.

polar coordinates an ordered pair consisting of a radius and an angle.

pole the fixed center of the polar coordinate system.

position vector another name for a standard vector.

positive angle results from counter-clockwise rotation.

primary solutions solutions defined over a limited domain.

principal nth root the unary root of a complex number.

product-sum identities useful in writing the product of trig functions as the sum and difference of trig functions.

projections another name for component vectors.

proving the identity showing the validity of one identity by using previously known facts.

Pythagorean identities fundamental identities that relate the sine and cosine functions and the Pythagorean Theorem.

quadrantal angle an angle in standard position with its terminal side on a coordinate axis.

quotient identities fundamental identities that involve the quotient of basic trig functions.

radian the measure on an angle with vertex at the center of a circle that subtends an arc equal to the radius of the circle.

radius vector another name for a standard vector.

real axis an axis in the complex plane.

reciprocal identities fundamental identities that involve the reciprocals of basic trig functions.

reduction formulas for cosine useful in rewriting cosines of angles greater than 90° as functions of acute angles.

reduction formulas for sine useful in rewriting sines of angles greater than 90° as functions of acute angles.

reduction formulas for tangent useful in rewriting tangents greater than 90° as functions of acute angles.

reference angle an acute angle whose trigonometric ratios are the same (except for sign) as the given angle.

resultant vector the result obtained after vector manipulation.

SAS reference to solving a triangle given the lengths of two sides and the measure of the included angle.

scalar multiplication changing the magnitude of a vector without changing its direction.

scalar multiplication of algebraic vectors a process of multiplying vector components.

scalar quantity the value of a dot product of two vectors.

secant the reciprocal of the cosine function.

second an angle measurement equal to 1/60 of a minute.

sector a portion of a circle enclosed by a central angle and its subtended arc.

semiperimeter one-half the perimeter of a triangle.

similar triangles two triangles whose angle measurements are the same.

simple harmonic motion a component of uniform circular motion.

sine a trigonometric ratio equal to the opposite side divided by the hypotenuse.

solving the triangle a process for finding the values of sides and angles of a triangle given the values of the remaining sides and angles.

SSA reference to solving a triangle given the lengths of two sides and the measure of a non-included angle.

SSS reference to solving a triangle given the lengths of the three sides.

standard position (angle) an angle with its initial side on the positive x-axis and vertex at the origin.

standard position (vector) a vector that has been translated so that its initial point is at the origin.

standard vector a vector in standard position.

static equilibrium the sum of all the force vectors add up to zero.

sum identities for tangent identities involving the tangents of sums of angles.

sum identity for cosine one of the trigonometric addition identities.

sum identity for sine one of the trigonometric addition identities.

sum-product identities useful in writing the sum and difference of trig functions as the product of trig functions.

tangent a trigonometric ratio equal to the opposite side divided by the adjacent side.

terminal point the ending point of a vector.

terminal side side of angle where angle measurement ends.

tip-tail rule a process for doing vector addition.

trigonometric addition identities identities involving the trig functions of sums and differences of angles.

trigonometric identity an equation made up of trigonometric functions of an angle that is valid for all values of the angle.

trigonometric ratios the ratios of the length of two side of a right triangle.

uniform circular motion circular motion about a point at a uniform linear and angular velocity.

unit circle a circle with a radius of one unit.

vector addition process of combining two vectors.

vector quantity a quantity that has both size and direction.

velocity vector a vector representing the speed and direction of a moving object.

vertical shift the vertical displacement of a function above or below the horizontal axis.

zero algebraic vector an algebraic vector whose components are both zero.

zero vector a vector with a magnitude of zero and any direction.

Index

Notes

Notes

Notes

Notes

Notes

Notes